Analog Circuits and Signal Processing

Series Editors
Mohammed Ismail
Mohamad Sawan

For further volumes:
http://www.springer.com/series/7381

Filip Tavernier • Michiel Steyaert

High-Speed Optical Receivers with Integrated Photodiode in Nanoscale CMOS

Springer

Dr. Filip Tavernier
K.U.Leuven
ESAT-MICAS
Kasteelpark Arenberg 10
B-3001 Heverlee
Belgium
filip.tavernier@esat.kuleuven.be

Prof. Dr. Michiel Steyaert
K.U.Leuven
ESAT-MICAS
Kasteelpark Arenberg 10
B-3001 Heverlee
Belgium
michiel.steyaert@esat.kuleuven.be

ISBN 978-1-4419-9924-5 e-ISBN 978-1-4419-9925-2
DOI 10.1007/978-1-4419-9925-2
Springer New York Dordrecht Heidelberg London

Library of Congress Control Number: 2011931863

Printed on acid-free paper

Springer is part of Springer Science+Business Media (www.springer.com)

Preface

These days, everything strongly relies on information. More and more, this information is digitized so that it can be stored and processed efficiently. The storage and processing cost is low and is still lowering thanks to the progress that is and has been made with the production technologies that are used for memories and processors, respectively. This evolution is caused mainly by the continuous scaling of transistors that can be realized on a chip. The consequence of this scaling is that a chip can contain an increasing number of transistors, a law which is generally known as *Moore's Law*. This ensures that more and more functionality can be integrated on a given area, which decreases the cost per function. However, the third cornerstone of information, namely communication, was never exposed to the same favourable development. By contrast with the processing and the storage of information, communication becomes increasingly critical because the applied channels cannot keep up with the high data rates.

Since long, optical communication is used as an alternative for wired and wireless communication networks because it offers a lot of advantages (high speed, low attenuation, low weight, high resistance against interfering signals, etc.), unfortunately at a high cost. Consequently, this type of communication is applied mainly in shared networks, such as transatlantic telephone networks, where the high cost is carried by the large number of users. In order to be able to use optical communication at a smaller scale, the expensive sender and receiver need to be manufactured in a standardized, cheap production technology. CMOS, the technology that is also used for the manufacturing of processors, is the preferred candidate. However, since CMOS is a standard process and is consequently not optimized to interact with light, the performance of the photodetector in this technology is weak compared to what can be achieved in dedicated processes.

The focus of the presented work is on the integration of the complete functionality of an optical receiver on a single CMOS chip. Due to the low speed and sensitivity of an integrated photodiode in this technology, innovative solutions are required both at the circuit and at the layout level. Therefore, a new photodiode is presented in this book that combines the speed of a differential photodiode and the sensitivity of a regular photodiode. The inherent disadvantage of a differential photodiode, namely the fact that half of the impinging light is reflected, is avoided with this *speed-enhanced photodiode*. The innovations on the circuit level can be found in the transimpedance amplifier as well as in the equalizer and the post amplifier. For example, to reduce the effect of the parasitic capacitance of the photodiode on the bandwidth of the transimpedance amplifier, the *capacitance-relieved TIA* is introduced. This topology enables the bandwidth of the circuit to be increased significantly without boosting the noise excessively. In addition, innovations are presented for the equalizer and the post amplifier which improve their performance considerably compared to the current state-of-the-art. This way, the problems of every building block of an optical receiver are dealt with in this work.

In order to demonstrate the functioning of the proposed concepts in a realistic situation, a number of chips is discussed. The first chip implements the functionality of an optical receiver in a 130 nm CMOS technology. The integrated differential photodiode has a diameter of 60 μm so that this receiver can be used together with a multi-mode fiber. Additionally, this chip has a differential transimpedance amplifier, a differential equalizer and a post amplifier. Signals with a maximum bit rate of 4.5 Gbit/s can be received while the bit error ratio is below 10^{-12}. The sensitivity equals −3.4 dBm. The total power consumption of this receiver is 74.2 mW, which is low compared to the current state-of-the-art.

The second presented optical receiver has a similar architecture and is implemented in the same 130 nm CMOS technology. However, a *speed-enhanced photodiode* is used instead of a differential photodiode. Consequently, the maximum bit rate rises to 5.5 Gbit/s, while the power consumption drops to 58.6 mW. As before, the sensitivity is −3.4 dBm.

The subsequent optical receiver discussed in this book has an integrated photodiode with a diameter of 1 mm, which is the same as the diameter of a plastic optical fiber. The related capacitance of this large photodiode in the used 180 nm CMOS technology equals 63.6 pF. This chip enables the reception of signals with a bit rate of 300 Mbit/s if the optical input power is between −14.5 and −9 dBm. Raising the supply voltage from the nominal 1.12 to 1.42 V, increases the maximum bit rate to 500 Mbit/s if the input signal has a power between −15 and −9.1 dBm.

The final chip that is discussed implements a post amplifier in a 90 nm CMOS technology. In order to realize the large time constant that is required by the offset compensation circuit, a new technique is used that boosts the capacitance of the integrated capacitor. Consequently, a larger time constant can be realized on a smaller area. The post amplifier delivers a gain of 35 dB and has a bandwidth of 4.15 GHz. The cut-off frequency due to the offset compensation circuit equals 1.86 MHz. The total power consumption of this chip is 14.7 mW, which is very low in comparison with the current state-of-the-art.

Contents

Acronyms and Symbols

AGC	Automatic Gain Control
ARC	Anti-Reflective Coating
BDP	Bandwidth-Distance Product
BER	Bit Error Ratio
BiCMOS	Bipolar CMOS
CATV	CAble TeleVision
CD	Compact Disc
CDR	Clock and Data Recovery
CHD	CHromatic Dispersion
CMOS	Complementary Metal-Oxide-Semiconductor
CW	Continuous Wave
DBR	Distributed Bragg Reflector
DCF	Dispersion Compensating Fiber
DDJ	Data-Dependent Jitter
DSL	Digital Subscriber Line
DVD	Digital Versatile Disc
EDGE	Enhanced Data Rates for GSM Evolution
ESA	European Space Agency
FOM	Figure Of Merit
FTTx	Fiber To The x
GPRS	General Packet Radio Service
GPS	Global Positioning System
GSM	Global System for Mobile Communications
HDTV	High-Definition TeleVision
IC	Integrated Circuit
ILD	Inter-Layer Dielectric
ISI	InterSymbol Interference
LED	Light-Emitting Diode
LFSR	Linear Feedback Shift Register
MD	Modal Dispersion
MESFET	MEtal-Semiconductor Field Effect Transistor
MMF	Multi-Mode Fiber

MOSFET	Metal-Oxide-Semiconductor Field Effect Transistor
MOST	Media Oriented Systems Transport
MSM	Metal-Semiconductor-Metal
NIC	Negative Impedance Converter
nMOS	N-channel MOSFET
NRZ	Non-Return-to-Zero
PCB	Printed Circuit Board
PDF	Probability Density Function
PIN	P-Intrinsic-N diode
PMD	Polarization-Mode Dispersion
PMMA	PolyMethyl MethAcrylate
pMOS	P-channel MOSFET
POF	Plastic Optical Fiber
PRBS	Pseudo-Random Binary Sequence
PSD	Power Spectral Density
RJ	Random Jitter
RMS	Root Mean Square
SATA	Serial Advanced Technology Attachment
SCR	Space Charge Region
SFP	Small Form-factor Pluggable
SMF	Single-Mode Fiber
SNR	Signal-to-Noise Ratio
SOI	Silicon-On-Insulator
TIA	TransImpedance Amplifier
UHF	Ultra High Frequency
UMTS	Universal Mobile Telecommunications System
UTP	Unshielded Twisted Pair
VCSEL	Vertical-Cavity Surface-Emitting Laser
VGA	Variable Gain Amplifier
WLAN	Wireless Local Area Network
XOR	eXclusive OR

BW	bandwidth of a circuit
c	speed of an electromagnetic wave in vacuum
C_{db}	drain-bulk capacitance of a MOSFET
C_{ds}	drain-source capacitance of a MOSFET
C_{gd}	gate-drain capacitance of a MOSFET
C_{gs}	gate-source capacitance of a MOSFET
C_{ox}	gate oxide capacitance per unit area of a MOSFET
C_{pd}	photodiode capacitance
C_{sb}	source-bulk capacitance of a MOSFET
D	dispersion parameter
D_{e^-}	electron diffusion constant
D_{h^+}	hole diffusion constant
f_t	transistor unity current gain frequency

GBW	gain-bandwidth product of a circuit
g_{ds}	small-signal admittance of the channel of a MOSFET
g_m	small-signal transconductance of a MOSFET
h	Planck's constant
I_{ds}	drain-source current of a MOSFET
k	Boltzmann's constant
L_{e^-}	electron diffusion length
L_{h^+}	hole diffusion length
PP	power penalty
q	elementary charge
r_{ds}	small-signal resistance of the channel of a MOSFET
r_e	extinction ratio
R_{pd}	photodiode responsivity
$R_\square$	resistance per square
T	temperature
T_b	bit period
t_{gd}	group delay
t_{pd}	phase delay
V_t	decision threshold
V_{th}	threshold voltage of a MOSFET
$V_{n,RMS}$	RMS noise voltage
Z_{TIA}	transimpedance of a TIA
ZBW	transimpedance-bandwidth product of a TIA
$\delta(x)$	impulse (Dirac delta function)
ε	permittivity
λ	wavelength
μ	mobility
μ_{e^-}	electron mobility
μ_{h^+}	hole mobility
ν	frequency
τ	time constant
τ_{e^-}	electron lifetime
τ_{h^+}	hole lifetime

Chapter 1
Introduction

1.1 Why Do We Need Optical Communication?

1.1.1 Existing and Emerging Applications

People work with information. All the time. In everything they do. To be useful, this information needs to be transported. This transportation of information, known as communication, can adopt several forms: making a telephone call, reading a newspaper, chatting with friends, downloading music onto your computer, etc. Nowadays, a growing amount of information is stored and transported in a digital format. Compared to analog data, digital data is easier to store, process and transport. This explains the continuing quest for faster, cheaper and more power efficient digital communication systems. Three important types of communication channels will be considered in this work: wireline channels, wireless channels and optical channels.

Most digital communication systems make use of a conducting connection, or wireline channel, between the sender and the receiver to transport data. One of the most typical examples in this category is the *unshielded twisted pair* (UTP) which is a copper wire connection that is established underground in a lot of the more developed regions of the world. This copper pair was originally intended to route telephone calls. However, with the rise of the internet in the last decade of the twentieth century, it was more and more applied to transport digital data. This evolution has led to the development of the *digital subscriber line* (DSL) technology which provides fast internet access from the central office to the subscriber, a connection which is also known as the *last mile*[1]. Thanks to this technology, a subscriber can make a telephone call and download and upload data all at the same time over a single UTP. Another example of a wireline communication system is *cable television* (CATV). Whereas this coaxial cable system was originally intended to distribute television signals, it is nowadays also used to provide the last mile access to the internet. In contrast to

[1] The connection from a communications provider to a customer is realized in different layers. High capacity and long distance connections are needed at the provider side whereas lower capacity and shorter distances are needed near the subscriber.

F. Tavernier, M. Steyaert, *High-Speed Optical Receivers with Integrated Photodiode in Nanoscale CMOS,* Analog Circuits and Signal Processing 5,
DOI 10.1007/978-1-4419-9925-2_1, © Springer Science+Business Media, LLC 2011

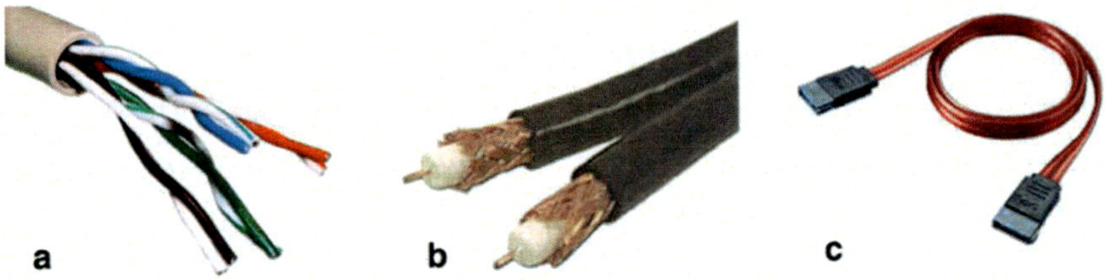

Fig. 1.1 Three examples of wireline communication channels. **a** UTP. **b** coaxial cable. **c** SATA cable

the DSL technology however, it combines the distribution of television signals and broadband internet access. In order to increase the quality of the delivered services and to enable the introduction of ***high-definition television*** (HDTV), television signals are these days more and more transmitted in a digital format. The CATV system is thus completely transformed from a dedicated analog medium into a shared digital one. A third example of an omnipresent wireline communication system can be found in every contemporary personal computer: SATA[2]. After its latest revision in 2008, SATA enables the transfer of data at a bit rate of 6 Gbit/s. The maximum supported distance is limited to 1 m. In Fig. 1.1 an UTP, a coaxial cable and a SATA cable are shown.

Where wireline systems excel in performance and reliability, wireless communication systems stand out in mobility. Because the sender and the receiver are not physically connected to each other in a wireless system, both can move freely without breaking the connection. Obviously, this is only true as long as the electromagnetic wave that is used to transport the data is strong enough at the location of the receiver. Without any doubt, the GSM[3] system is the most widespread digital wireless communication system in the world. This second generation (2G) mobile telephone network differs from its first generation (1G) counterpart in the fact that the data is digitized before being transmitted. The GSM standard prescribes the transmission of digital data at a rate of 10 kbit/s which is enough to make a phone call. Streaming video and basic web applications became available on mobile devices with the introduction of 2.5G mobile networks (GPRS, EDGE) in the year 2001. The introduction of 3G mobile networks (UMTS) in 2004 further increased the possibilities, especially in terms of data rate. The *global positioning system* (GPS) is also an application that uses a wireless channel. This navigation system needs at least 24 satellites in a fixed orbit around the earth to be operational. If the signal from at least four of those satellites is received properly, the geographical position of the receiver can be retrieved with an accuracy of about 10 m. The Galileo navigation system, which is the European counterpart of the American and military inspired GPS, is currently under deployment. It will provide an accuracy of several millimeters and a better coverage

2 Serial Advanced Technology Attachment (SATA) is a serial computer bus interface which is used to transport data between the processor and the hard disk.

3 GSM stands for Global System for Mobile Communications, or formerly, Groupe Spéciale Mobile.

for countries with a higher geographical latitude. A final example in this category is the *wireless local area network* (WLAN). The most important standard in this field is IEEE 802.11 which is generally denoted as Wi-Fi. Traditionally, Wi-Fi is used to distribute internet inside houses and other buildings. Nowadays, it is also applied to provide internet access to complete cities, especially in developing regions that do not have sufficient UTPs or coaxial cables available at the curb. Compared to 2.5G and 3G mobile networks, Wi-Fi enables higher data rates but at the expense of reduced mobility.

Surprisingly or not, optical communication is the oldest communication technique among the ones that are discussed in this section. Body language, beacon fires or smoke signals can all be considered to be the ancestors of the modern communication system as it is perceived today. The light that carries the useful data can be transmitted in two fundamentally different manners, resulting in free-space optical communication and optical fiber communication, respectively. The former technique can be used for communication between spacecraft. For example, in 2006 the *European space agency* (ESA) realized a laser link of 40, 000 km between its ARTEMIS satellite and an aircraft, an achievement equivalent to targeting a golf ball over the distance between Paris and Brussels [ESA10]. This is shown conceptually in Fig. 1.2. Instead of using an UTP or coaxial cable to bridge the last mile for telecommunications, an optical fiber can be applied also. This technology is generically gathered under the term *fiber to the x* (FTTx). *Fiber to the home* (FTTH) is a form of the FTTx technology where a fiber connection is deployed between the central office and every house. Other possibilities are FTTB (building), FTTC (cabinet) and FTTN (node). In the case of these last three configurations, a copper connection is applied between the building (FTTB), the cabinet (FTTC) or the node (FTTN) and the subscriber to complete the link. The main advantage of this technology compared to a more traditional wireline last mile solution is the higher speed that can be obtained and the longer distance that can be bridged by an optical fiber. Another application domain of optical communication is the transportation of data between two or more processor cores on a single chip or between two processor chips. Electrical connections that

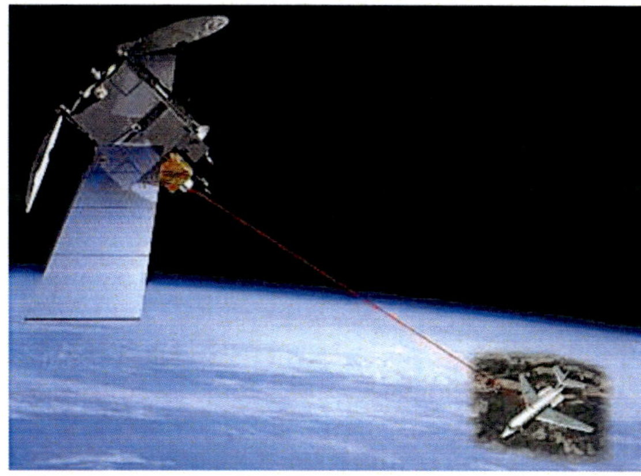

Fig. 1.2 Impression of a laser data link between Artemis and an aircraft [ESA10]

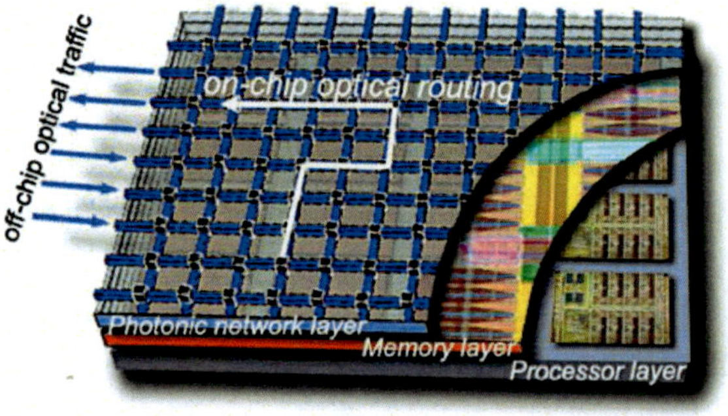

Fig. 1.3 Concept of a 3D silicon processor chip with an optical routing layer [IBM10]

are traditionally used to perform this task suffer from severe bandwidth limitations and a corresponding high power consumption. In [IBM10], optical interconnects are proposed to replace such electrical connections. This is shown conceptually in Fig. 1.3. As it already happened in long communication links some decades ago when optical fibers replaced electrical connections, the UTPs and coaxial cables that provide the last mile connection are now starting to be replaced by optical fibers. Following this trend, optics will become competitive with copper for increasingly shorter links, eventually leading to optical inter-chip and intra-chip communications.

1.1.2 The Advantages of Optical Communication

Due to the nature of its channel, optical fiber communication has a number of significant advantages compared to wireline or wireless communication links. The most important of these are listed below.

- The **bandwidth** of an optical fiber is extremely high. This means that data can be transmitted at an accordingly high data rate. Because the bandwidth of a fiber decreases linearly with length, a fiber is often characterized by its *bandwidth-distance product* (BDP). A typical multi-mode fiber has a BDP of 2 GHz × km for a wavelength of 850 nm. This means that a 2 GHz signal can be transmitted over a fiber length of 1 km while the attenuation is only −3 dB. For comparison, a Cat-6 UTP has a BDP of only 20 MHz × km.
- The **loss** of an optical fiber is extremely low. This means that an optical signal can be transmitted over a very long distance before needing a signal repeater. The loss of a silica fiber is only 2.5 dB/km for a wavelength of 850 nm. Around a wavelength of 1.55 μm the loss is even lower at 0.25 dB/km. Therefore, this longer wavelength is typically used for long-haul communication systems whereas a wavelength of 850 nm is used for short-range data communication. These loss values are phenomenally low compared with the losses in electrical channels. As a reference, a high-performance coaxial cable has a typical loss of 500 dB/km for a signal at 10 GHz. This means that the power of a 10 GHz signal

is halved after 6 m of coaxial cable whereas this is only the case after 1.2 km of silica fiber at a wavelength of 850 nm and after 12 km of fiber when a wavelength of 1.55 μm is used.

- Due to its non-metallic character, an optical fiber is not susceptible to **electromagnetic interference**. It is therefore an ideal medium to be applied along railroad tracks, utility lines or close to a combustion engine in automobiles.
- Due to its non-metallic character, an optical fiber is not susceptible to **crosstalk** because there is no conductive, capacitive or inductive coupling to neighboring communication channels. For that reason, signal integrity is always maintained in an optical fiber.
- The lack of electromagnetic emission makes an optical fiber a very **secure** channel. It is almost impossible for a non-authorized person to retrieve the data that is sent which makes it a perfect candidate in military applications. This is in strong contrast with a wireless connection. The air is by definition a shared medium so everyone can retrieve the transmitted data without being noticed by the sender or the receiver. Consequently, data that is transmitted over a wireless channel needs always to be encrypted in order to guarantee some form of security.
- The sender and receiver of an optical communication system are **galvanically separated** from each other. Especially for long-haul communication links this property may not be underestimated. If a conducting cable is used to communicate, a different ground potential between the sender and the receiver can cause significant currents to flow in the connecting ground wire which in turn can introduce an offset voltage and a degradation of the sensitivity.
- The **physical size and weight** of an optical fiber is very low compared to its electrical counterpart. An optical fiber with its protective jacket has a typical diameter of 0.635 cm. Nevertheless it can replace a bundle of copper wires with a diameter of 7.62 cm to carry the same number of telephone conversations or other signals. Obviously, the weight is reduced accordingly when a fiber is installed instead of using a bundle of copper wires. 94.5 kg of copper wire can thus be replaced by 3.6 kg of optical fiber. This property is invaluable for applications in aviation and space travel.
- The **cost** of an optical fiber is lower than that of a bundle of copper wires with the same transmission capacity.

Although it is stated that the cost of an optical fiber is lower than the equivalent copper wire capacity, the cost of a complete optical communication system is significantly higher than its electrical counterpart. The reason for this is the expensive components used to manufacture the optical transmitter and receiver. This cost issue is the most important reason why optical communication systems cannot yet compete with the more traditional electrical solutions in low-cost applications for the consumer market. The development of completely integrated optical receivers in low-cost technologies, the subject of this thesis, can therefore be seen as the last hurdle for optical communication.

1.2 What are Integrated Optical Receivers?

An electronic network or circuit is a combination of electronic components that performs a useful function. This functionality can adopt several forms: a microprocessor, a radio, a power amplifier, an optical receiver, etc. Several manufacturing processes and technologies can be applied to realize this combination of electronic components.

A *printed circuit board* (PCB) is a board with copper interconnection lines that is used as a rigid base to support the discrete electronic components of which a circuit is composed. Resistors, capacitors, inductors, transistors, etc. can be connected to each other reliably in a certain circuit configuration. An example of a PCB is shown in Fig. 1.4. The holes to attach the discrete components and the interconnection lines to connect them can easily be noticed.

On the contrary, an *integrated circuit* (IC) is a miniaturized electronic circuit that is realized on a single piece of semiconductor material. It integrates the functionality of a PCB with its associated discrete components on a single piece of material and is therefore a very area-efficient way to manufacture a circuit. This technology became possible after the invention of the semiconductor transistor in 1947 by John Bardeen and Walter Brattain. From then on, bulky and power hungry vacuum tubes could be replaced by much smaller and power efficient transistors. The invention of the IC in 1958 by Jack Kilby enabled the integration of an enormous number of very tiny components, especially transistors, on a small chip. These components are all manufactured on the same substrate using photolithography. Due to its small size and vulnerable connections (bondwires), an IC is always packaged. An example is shown in Fig. 1.5.

Since its invention in 1958, IC technology has evolved tremendously. This evolution is described by Moore's law which states that the number of transistors that can be realized inexpensively on an IC doubles approximately every two years [Moo65]. This exponential growth is already going on for more than four decades. The result

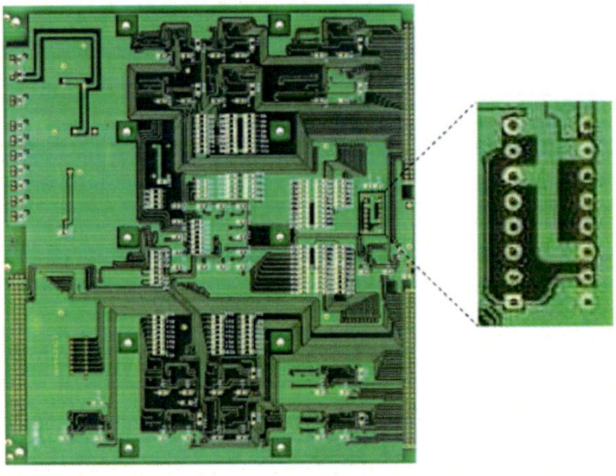

Fig. 1.4 Example of a PCB: components are soldered on the board and are connected by the etched copper traces

Fig. 1.5 Photo of a chip in a package. Bondwires connect the chip terminals with the terminals of the package

of this growth is that contemporary ICs have a transistor count that is more than a million times higher than that of ICs around 1970. This is made possible by the continuing device scaling. In 1970, a typical transistor had a gate length of several micrometers whereas a transistors nowadays can have a gate length of only 32 nm. This scaling has resulted in a tendency to integrate more and more functionality on a chip which results in a lower cost price, lower power consumption and higher performance.

Complementary metal-oxide-semiconductor (CMOS) is the dominant technology to manufacture ICs. It allows the fabrication of ***metal-oxide-semiconductor field effect transistors*** (MOSFETs) with an n-type channel (nMOS) and a p-type channel (pMOS) on one and the same chip. A cross section of a typical CMOS process is depicted in Fig. 1.6. In most cases a p-type substrate is used with a relatively low impurity or doping concentration. An nMOS is formed inside a p-well which is a local region in the substrate having a higher doping concentration than the p-type substrate. This p-well forms the bulk of the nMOS. The bulk contact (B) is realized by a very highly doped p-type region (p+). The source (S) and drain (D) are formed by highly doped n-type regions (n+). The gate (G) is placed on top of the substrate. To shield the gate electrically from the channel underneath it, an isolating layer is placed

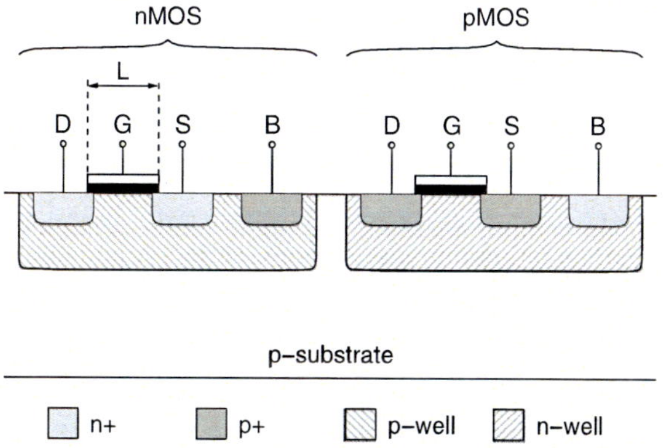

Fig. 1.6 Cross section of a CMOS process with a p-substrate

in between. In recent CMOS technologies the gate material is polysilicon while the insulating layer is a form of silicon dioxide. By contrast, a pMOS is constructed in an n-well. The bulk connection is provided by a highly doped n-type region whereas the source and drain are formed by highly doped p-type regions. The gate of a pMOS has an identical construction to that of an nMOS. The behavior of a transistor is determined by its length (L) and width. The minimal length that can be realized in a certain technology is one of the most important scaling parameters.

As mentioned before, recent CMOS technologies are able to produce devices with a minimal length of only 32 nm. The quest for even smaller devices continues relentlessly because scaling has a lot of beneficial effects. The most obvious advantage is the fact that scaling allows one to pack more devices on the same chip area. As a result, the same functionality can be realized in a smaller area or, equivalently, in the same area more functionality can be integrated. Because the cost of a chip is predominantly determined by its area, the cost of a certain functionality is reduced thanks to scaling. Secondly, scaling makes the devices switch faster. This is due to the fact that, in order to conduct a certain current, the width of a transistor can be decreased as the length is reduced. In conclusion, the device area and the accompanying parasitic capacitances are decreased accordingly, enabling higher operating speeds.

Whereas CMOS was originally intended to integrate purely digital circuits, more and more analog functionality is realized with it nowadays. The most important motive is the cost of CMOS technologies that is very low compared to the more dedicated processes that were typically used to manufacture analog circuits. Integrating a complete optical receiver in CMOS, the subject of this thesis, can therefore be justified on economic grounds alone.

1.3 Overview of Existing Literature

The field of optical communication, and that of optical receivers in particular, is treated extensively in open literature. The objective of this section is to give an overview of the most recent developments in the field. First, the implementations are discussed that are not integrated in a standard CMOS technology. Next, some CMOS realizations are shown that make use of an external photodiode. Finally, the CMOS implementations with an integrated photodiode are treated.

1.3.1 Non-CMOS Implementations

Non-CMOS implementations represent the major portion of recently published optical receivers. This is due to the fact that the performance is maximized if the optimal technology is chosen for every building block. For the photodiode, it is important to have a high conversion factor from light power to electrical current. Moreover, this conversion should be performed as fast as possible. It is also beneficial for the photodiode to exhibit a minimal parasitic capacitance. Typically, these demands lead to a photodiode that is not integrated in a silicon CMOS process.

In [You10], a photonic CMOS process is proposed to establish a complete optical link at chip level. This process enables the integration of light modulators, optical waveguides and optical detectors at the same time. The detector, for example, is realized by means of a *metal-semiconductor-metal* (MSM) junction with germanium as semiconductor material. The waveguides make use of a silicon nitride film that is surrounded by silicon dioxide. In [Ana06, Nar07], a comparable photonic process is presented. The 0.5 μm wide optical waveguides in this case are realized with silicon surrounded by silicon dioxide. Additionally, a holographic lens, consisting of trenches etched into the active silicon, is used as an optical detector. The trenches scatter light from a normally incident optical fiber into the silicon waveguide. Obviously, these photonic processes differ immensely from the commonly used standardized semiconductor processes. These techniques can therefore only be seen as non-standard solutions that might have value in the long-term.

The receiver that is presented in [Hag91] makes use of a GaAs substrate. The available *metal-semiconductor field effect transistors* (MESFETs) enable the development of a receiver that can receive data at a bit rate of 10 Gbit/s. On the contrary, the presented receiver in [Hua04] is implemented in a 0.35 μm SiGe BiCMOS process. Again, a bit rate of 10 Gbit/s can be processed reliably. A 0.35 μm SiGe BiCMOS process is also used in [Che04]. The maximum bit rate is only 1.25 Gbit/s, while the receiver has a very high dynamic range and is able to operate in a burst-mode application. The photodetector is an external component in all these implementations.

The majority of realizations use a process with a silicon substrate. Two main categories can be distinguished: *silicon-on-insulator* (SOI) CMOS and silicon bipolar. An implementation in a 130 nm SOI CMOS process is presented in [Csu02]. The integrated photodiode is biased with a reverse voltage of 20 V which is considerably higher than the supply voltage of the receiver. Thanks to this fact, and to the almost ideal substrate shielding of the process, the receiver is able to operate at a bit rate of 8 Gbit/s.

In the silicon bipolar receivers that are presented in the literature, the photodetector is typically integrated along with the circuitry because it is possible in these processes to build a PIN photodiode[4] wherein the n-type and p-type regions are well separated from each other. It will be shown in Chap. 3 that this leads to some very important advantages compared to a regular PN junction. Optical receivers with an integrated photodiode are presented in [Tak98, Stu05]; these are intended to be operated in an optical storage application. The receiver in [Tak98] has a bandwidth of only 56 MHz and can be used as an optical pickup unit for a CD-ROM. On the contrary, the pickup unit that is proposed in [Stu05] has a bandwidth of 260 MHz which is high enough to read a Blu-ray disc. The chip photograph of this integrated pickup unit is shown in Fig. 1.7. In [Tad08], an implementation in a 0.6 μm BiCMOS process is

[4] A PIN diode is a diode with a lightly doped, almost intrinsic, semiconductor region between the p-type semiconductor and the n-type semiconductor regions.

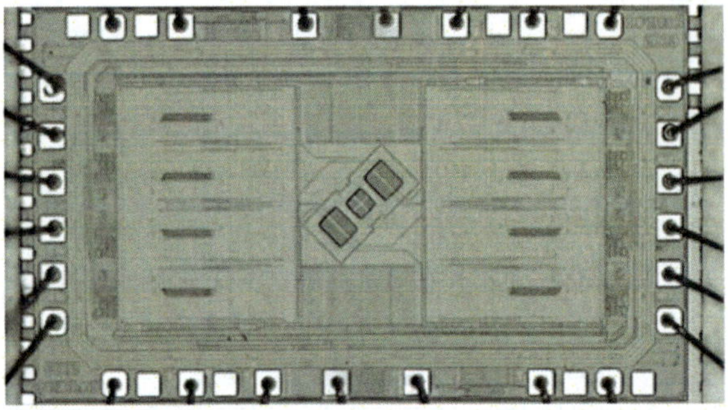

Fig. 1.7 Chip photograph of the optical receiver chip for CD/DVD/Blu-ray disc applications [Stu05]

presented with an integrated PIN photodiode. The achieved bandwidth is 205.5 MHz. In [Swo06], a bit rate of 11 Gbit/s is attained. A 0.5 μm BiCMOS process is used in which a PIN photodiode with a diameter of 50 μm is integrated. The receiver that is discussed in [För06] is integrated in a 0.6 μm BiCMOS process. The maximum bit rate is only 300 Mbit/s but the integrated photodiode has a huge diameter of 500 μm.

1.3.2 CMOS Implementations Without an Integrated Photodiode

Another major portion of recently published optical receivers are those that are implemented in a standard CMOS technology but without an integrated photodetector. These implementations only provide the conversion from input current to output voltage. The conversion of the light signal to the electrical domain needs to be performed by an external component.

To measure a receiver without an integrated photodetector, one can typically follow two procedures. The first one is to simulate the behavior of a real photodiode by building its equivalent model with off-chip components. As a photodiode acts like a modulated current source, it can be replaced by its Thévenin equivalent network[5]. Therefore, its bbehaviorcan be simulated by a voltage source and a high-ohmic resistor in series. The photodiode capacitance is modeled by a separate capacitor having the same capacitance as the photodiode. This procedure is followed to measure the receiver that is proposed in [Ing99]. In a 0.7 μm CMOS process, a combination of a transimpedance amplifier and limiting amplifier is realized that is able to operate at a bit rate of 1 Gbit/s. The chip photograph is shown in Fig. 1.8. In [Che06] the receiver is integrated in a newer technology with a minimal linewidth of 0.35 μm. The maximum bit rate for this implementation is 2.5 Gbit/s. Thus, a technology with

[5] Thévenin's circuit theorem states that any combination of voltage sources, current sources and resistors with two terminals can be replaced by a single voltage source and a single series resistor. Norton's circuit theorem, on the contrary, states that one can replace these type of networks by a single current source and a single parallel resistor.

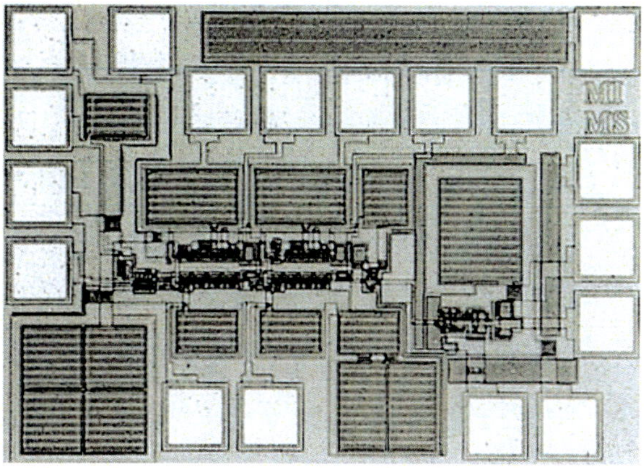

Fig. 1.8 Chip photograph of the 1 Gbit/s optical receiver chip [Ing99]

a smaller minimal linewidth seems to be advantageous to increase the maximum bit rate. The same observation can be made for the receiver with a bandwidth of 5.3 GHz in [Pis05] which is integrated in a 0.18 µm CMOS process. A final implementation which is measured with a simulated photodiode is presented in [Tan98]. A maximum bit rate of 2.4 Gbit/s is reached in a 0.15 µm CMOS process.

The second procedure to measure an optical receiver without an integrated photodiode is actually the more precise one: connecting an external photodiode at the input. Doing so, all parasitic effects are taken into account and the measurements are therefore a better reflection of the actual operation in an application. In [Sch02], a 1.25 Gbit/s optical receiver is proposed which is integrated in a 0.35 µm CMOS process. The authors do not disclose any information about the photodiode utilized to do the measurements. In [Che05], an external photodiode is used with a responsivity of 0.85 A/W and a capacitance of only 0.15 pF. Due to this very small capacitance the presented receiver, which is implemented in 0.18 µm CMOS, is able to operate at a bit rate of 10 Gbit/s. The implementation in [Che07c] which is realized in the same technology reaches an identical maximum bit rate. A CMOS process with the same minimal linewidth is used for the receivers that are presented in [Yun09, Che09]. The former operates at a maximum bit rate of 4 Gbit/s whereas 2.5 Gbit/s is the maximum bit rate for the latter. In [Azu08], a realization is proposed that is composed of two separate chips. The first chip contains the transimpedance amplifier whereas the second chip incorporates a limiting amplifier and clock and data recovery circuit. Both are integrated in a standard 0.13 µm CMOS process. The attained bit rate is 1.25 Gbit/s. This receiver is meant to be used in mobile handsets. The same technology is used for the receiver that is discussed in [Sch09]. However, the maximum bit rate is significantly higher, namely 10 Gbit/s. A CMOS process with a minimal linewidth of 80 nm is used for the realization that is proposed in [Kro05]. The reported maximum bit rate is again 10 Gbit/s. Finally, in [Pal08] an implementation in 90 nm CMOS is proposed which is able to operate at 16 Gbit/s. To obtain such a high bit rate, an integrating optical receiver is used instead of the conventional amplifying architecture. The drawback of this choice is that one needs a reliable synchronous clock signal in the receiver which is not always available.

1.3.3 CMOS Implementations with an Integrated Photodiode

The final category of optical receivers are the ones that are completely integrated in CMOS. They provide the conversion from light to output voltage in a single chip. Obviously, this is very interesting from a cost and reliability point of view. Because a standard CMOS process is not at all adapted to integrate light sensitive devices, there are some issues that need to be tackled first. Although these are discussed in Chap. 3, it can already be noticed here that the implementations in CMOS with an integrated photodiode typically operate at a lower bit rate than the ones that were discussed in the previous sections. This is due to the low conversion speed from light to electrical current of a CMOS integrated photodiode.

In the literature, this low conversion speed is typically circumvented in three different ways: applying a reverse voltage over the photodiode that is higher than the supply voltage, using a differential photodiode topology or adding an optimized equalizer. In [Woo98], a receiver is presented which is integrated in a 0.35 μm CMOS technology. The photodiode with an area of only 16.54 μm × 16.54 μm is reverse biased with a voltage of 10 V whereas the supply voltage is only 3.3 V. The attained bit rate for light with a wavelength of 850 nm is 1 Gbit/s. The implementation in [Che07a] is integrated in a 0.18 μm CMOS process. A reverse voltage of 6 V is applied at the photodiode to increase its speed. This results in a maximum bit rate of 2.5 Gbit/s.

The differential photodiode topology was introduced for the first time in [Roo00]. The receiver that uses a 50 μm × 50 μm photodiode having this topology attains a bit rate of 250 Mbit/s. In [Jut05], an optical receiver was introduced that also uses this type of photodiode. Additionally, the photodiode is biased with a voltage of 3.3 V which is higher than the nominal supply voltage of the used 0.18 μm CMOS process. The achieved bit rate is 2 Gbit/s.

The third option to increase the speed of an integrated photodiode is by adding an equalizer in the receiver chain. This is done in [Rad05], where a realization in 0.18 μm CMOS is presented that achieves a bit rate of 3 Gbit/s. Without the equalizer, the bit rate is below 50 Mbit/s which is caused by the low photodiode bandwidth of only 10 MHz. This huge speed improvement is obtained by adapting the equalizer frequency response to that of the photodiode. The chip photograph of this realization can be seen in Fig. 1.9.

Naturally, the combination of the last two techniques potentially leads to even higher bit rates. This has been done for the receiver in [Che07b] which is integrated in a 0.18 μm CMOS process. The reported bit rate is 3.125 Gbit/s. A technology with the same minimal linewidth is used to integrate the receiver of [Kao09]. However, the maximum bit rate is significantly higher, namely 5 Gbit/s. In [Lee10], a receiver

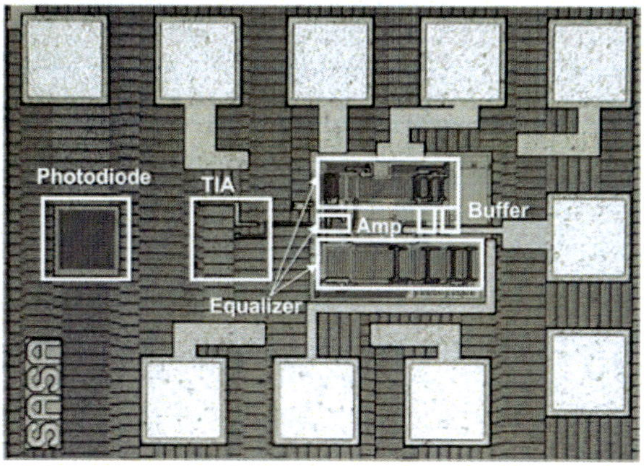

Fig. 1.9 Chip photograph of the completely integrated optical receiver [Rad05]

that achieves a bit rate of 8.5 Gbit/s is presented which is realized in a technology with a smaller minimal linewidth of 0.13 μm.

1.4 Summary of the Research

From the literature overview in Sect. 1.3, it can be noticed that the speed of optical receivers that are completely integrated in a standard CMOS process is generally lower than that of receivers where all or some of the building blocks are integrated in a more dedicated and thus more expensive technology. Also, the sensitivity of receivers with a CMOS integrated photodiode is not as high as what can be achieved in other technologies. On the contrary, from a cost and reliability point of view it is very advantageous to integrate all the functionality of an optical receiver on a single chip that is realized in a standard technology.

Therefore, the main objective of this research work was to assess the possibilities of standard CMOS technologies to serve as the production technology for high-speed optical receivers. To achieve this, several chip implementations are designed, processed and measured. All building blocks from an optical receiver are treated, except for the clock and data recovery system. New concepts and techniques are introduced to improve the overall performance of completely integrated CMOS optical receivers.

The main focus of the research was on increasing the speed of this kind of receivers. Obviously, there are numerous other specifications that determine the quality of such a system. Three of the most important are the sensitivity, the power consumption and the physical size of the system and the photodetector in particular. Therefore, these parameters are also considered extensively when new concepts and techniques are explored. The objective is to improve the overall trade-off between all relevant parameters.

1.5 Outline of the Thesis

As discussed in the previous section, the objective of the performed research was to develop, design and measure optical receivers. These optical receivers should have two important and renewing properties: they should be integrated in a standard nanoscale CMOS technology and, additionally, the light-sensitive component, the photodiode, should be integrated in the same low-cost unaltered process. In order to put things in a correct perspective, the theoretical background is treated first in this thesis. After that, some implementations and their respective measurements are discussed.

In Chap. 2, a high-level perspective is given on the subject of this thesis. It introduces the reader into the world of data communication and that of optical communication in particular. Some important concepts are introduced and the basic trade-offs are explained.

The photodiode performs the task of converting light into electricity. As it is the very first building block of an optical receiver it determines its performance to a great extent. Integrating the photodiode in a CMOS technology brings a number of particular issues which affect its conversion efficiency, speed and parasitics. All of these are discussed in Chap. 3.

The transimpedance amplifier converts the current that is generated by the photodiode into an amplified voltage. Especially the design of this building block is submitted to a lot of important trade-offs because it needs to combine high gain, high bandwidth, low equivalent input noise and low power consumption. A thorough discussion is given in Chap. 4.

One of the problems that arises while integrating the photodiode in a standard CMOS technology is the low conversion bandwidth from light to electricity which severely limits the attainable bit rate. Therefore, an equalizer is added to the receiver chain. This equalizer is treated in detail in Chap. 5.

The post amplifier boosts the signal swing to a value that is more or less as high as the supply voltage of the system. It needs to combine a high gain and a large bandwidth. An offset compensation circuit is typically applied to compensate for the disadvantageous effects of an offset voltage on the circuit performance. This building block is discussed in detail in Chap. 6.

In Chap. 7 an overview of the realized chip implementations is given. All of these realizations integrate one or more building blocks which are discussed in the other chapters. Measurements are shown to validate the theory. Additionally, a comparison is given with the current state-of-the-art.

Finally, the general conclusions are drawn in Chap. 8. On overview is also given on the main contributions of this work and some suggestions are made for further research.

Chapter 2
Optical Communication—A High-Level Perspective

As suggested in Chap. 1, the need for low-cost optical communication systems is omnipresent. Before discussing the design of an optical receiver in detail in the next chapters, it is important to understand the basic concepts of communication systems in general and the most significant aspects of an optical system in particular. An overview is given in this chapter.

2.1 The Communication Model

The objective of communication is to transport a message between two or more entities. One of these entities generates the data to be transported and is called the *transmitter*. A *channel* is needed to transport the message to the addressee, the *receiver*. This is shown schematically in Fig. 2.1. The transmitter consists of a driver and a light source. The driver conditions the digital data so that it can be applied to the light source, which is typically a *laser diode*. In this way, the signal is converted to the optical domain which is necessary to be guided through the optical fiber. At the other end of the fiber, the light signal is converted back to the electrical domain by a photodiode. However, the signal which is generated by the photodiode cannot be interpreted as digital data because typically it is greatly attenuated and distorted during transportation through the fiber. Therefore, the receiver needs to perform three operations before the data can be properly understood: *re-amplification*, *re-shaping* and *re-timing*. Because of this, such a receiver is sometimes called a *3R-receiver*.

A *transceiver* is the combination of a transmitter and a receiver in a single package. Mostly, such a transceiver is implemented in a *small form-factor pluggable* (SFP) format. An example is shown in Fig. 2.2. In this manner, a full-duplex communication link can be established by means of two half-duplex links in parallel.

The communication model of Fig. 2.1 can be used in a vast variety of application domains. Some of them were discussed in Chap. 1. Although they do not always fully comply with the presented model, the basic concept remains the same. For example, in an optical pick-up unit for a CD, DVD or Blu-ray disc, light is modulated by the disc which can therefore be considered as the transmitter. Moreover, the channel exists from air and a lens system instead of an optical fiber.

F. Tavernier, M. Steyaert, *High-Speed Optical Receivers with Integrated Photodiode in Nanoscale CMOS,* Analog Circuits and Signal Processing 5,
DOI 10.1007/978-1-4419-9925-2_2, © Springer Science+Business Media, LLC 2011

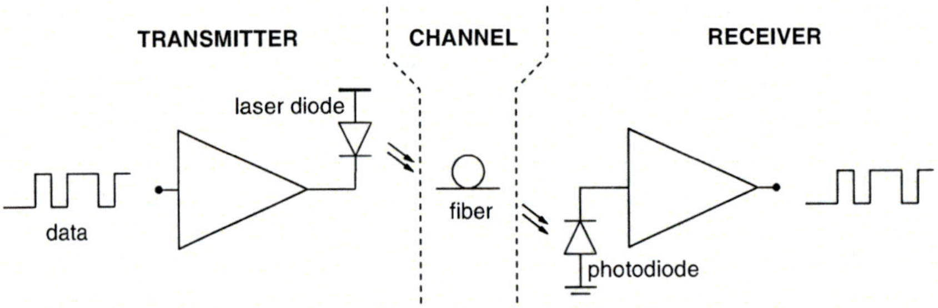

Fig. 2.1 A simple optical communication system, consisting of a transmitter, a channel and a receiver. The laser diode at the transmitter converts the electrical signal into modulated light while the photodiode at the receiver does the opposite

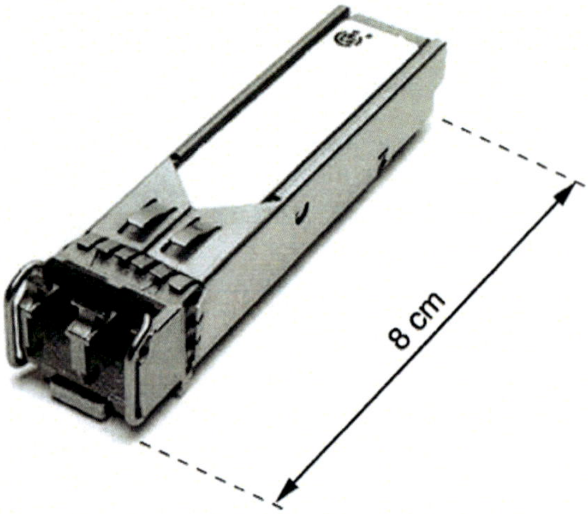

Fig. 2.2 An SFP optical transceiver; one fiber input is used to send data, the other fiber input is used to receive data

2.2 Properties of Binary Data

2.2.1 Random Binary Data

One of the common aspects for almost all optical communication systems is the type of data signals which are used. Logical ONEs and ZEROs typically modulate the amplitude of the transmitted lightwave. This will be clarified in Sect. 2.3. An example waveform of a binary data signal is shown in Fig. 2.3a. A high amplitude of the waveform corresponds with a logical ONE whereas a low amplitude corresponds with a logical ZERO. The duration of a single bit, the *bit period*, is represented by T_b. The reciprocal of T_b is called the *bit rate*. Since they are random in nature, a ONE and a ZERO occur typically with an equal probability. The waveform in Fig. 2.3a is

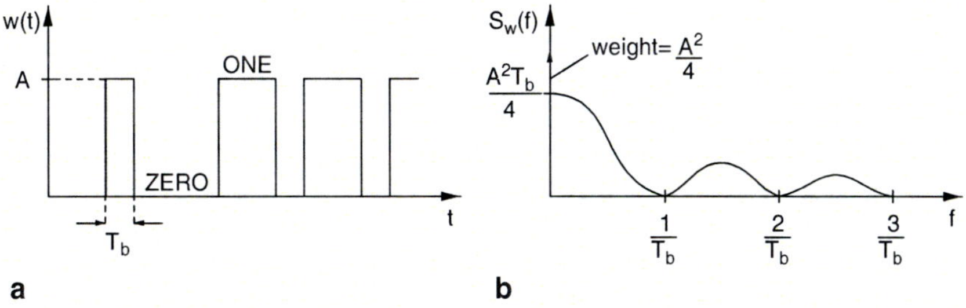

Fig. 2.3 Example of a logical binary signal and its accompanying spectrum. **a** Waveform. **b** Spectrum

mathematically described as:

$$w(t) = \sum_i b_i p(t - iT_b), \tag{2.1}$$

where $b_i = 0$ for a ZERO and $b_i = 1$ for a ONE. The pulse shape, $p(t)$, is defined as follows:

$$p(t) = \begin{cases} A & \text{if } 0 \leq t \leq T_b \\ 0 & \text{otherwise.} \end{cases} \tag{2.2}$$

The *power spectral density* (PSD) is normally used to analyze the frequency domain behavior of a signal. For a waveform that can be described with (2.1) and for a ONE and a ZERO having an equal probability to appear, the PSD is given by [Cou02]:

$$S_w(f) = \frac{|P(f)|^2}{4T_b}\left(1 + \frac{\delta(f)}{T_b}\right), \tag{2.3}$$

where $P(f)$ represents the Fourier transform of $p(t)$ and $\delta(f)$ is the *Dirac delta function*. If $p(t)$ is defined by (2.2), its Fourier transform is given by [Cou02]:

$$P(f) = AT_b \frac{\sin \pi f T_b}{\pi f T_b} e^{-j\pi f T_b}. \tag{2.4}$$

Using (2.4), (2.3) can be rewritten as:

$$S_w(f) = \frac{A^2 T_b}{4}\left(\frac{\sin \pi f T_b}{\pi f T_b}\right)^2 \left(1 + \frac{\delta(f)}{T_b}\right) \tag{2.5}$$

This spectral behavior is shown in Fig. 2.3b. Obviously, most of the signal power is contained in the frequency band below $f = 1/T_b$. Because the average of the data signal (2.1) is not zero, an impulse is present at DC. Interestingly, there is no spectral component at $1/T_b$, the bit rate of the signal. From now on the term *spectrum* is loosely used instead of PSD.

This type of data signal is known as *unipolar non-return-to-zero* (NRZ) signaling. It is unipolar because the signal is not symmetrical with respect to the zero level. This is actually the case when the data modulates the power of a lightwave: the power cannot drop below zero. As the amplitude remains constant during the entire bit period, the signal is said to be NRZ. A *return-to-zero* (RZ) waveform would return to the zero level for the second half of the bit interval. Compared to a NRZ signal, the spectrum of a RZ signal is contained in a frequency band that is twice as wide. This has an effect on the required bandwidth of the receiver which will be explained in Sect. 2.5. On the other hand, an impulse at $1/T_b$ is present in the spectrum of a RZ signal, which simplifies the re-timing task of a 3R-receiver which is performed by the *clock and data recovery* (CDR) system.

2.2.2 *Pseudo-random Binary Data*

In contrast to a truly random binary data stream, a *pseudo-random binary sequence* (PRBS) is completely predictable. Such a PRBS is typically produced by a *linear feedback shift register* (LFSR), of which an example is shown in Fig. 2.4a. An LFSR has a number of clocked memory elements and a linear feedback loop around (some of) them. The linear feedback is provided by means of the *exclusive OR* (XOR) gate which is in fact the only linear function of single bits.

The topology of Fig. 2.4a passes cyclically through a predefined number of states. Suppose the initial state is 111. At the next clock cycle, this initial state changes to 011. The succession of states for this topology is represented in Fig. 2.4b. After the last state, the circuit returns to its initial state and the process starts all over again.

The signal X_1 forms a PRBS as do X_2 and X_3, which are delayed versions of X_1. The length of the generated pattern is determined by the number of memory elements in the LFSR. For three memory elements, as is the case in Fig. 2.4, this

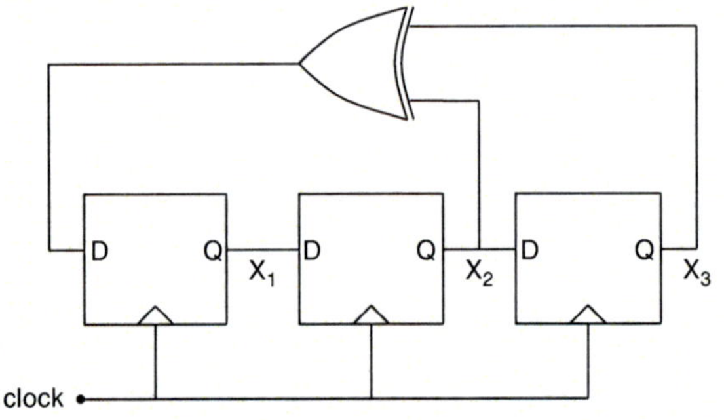

state	X1	X2	X3
1	1	1	1
2	0	1	1
3	0	0	1
4	1	0	0
5	0	1	0
6	1	0	1
7	1	1	0

return to state 1

b

Fig. 2.4 Generation of a PRBS with a length of $2^3 - 1$. **a** LFSR. **b** State table

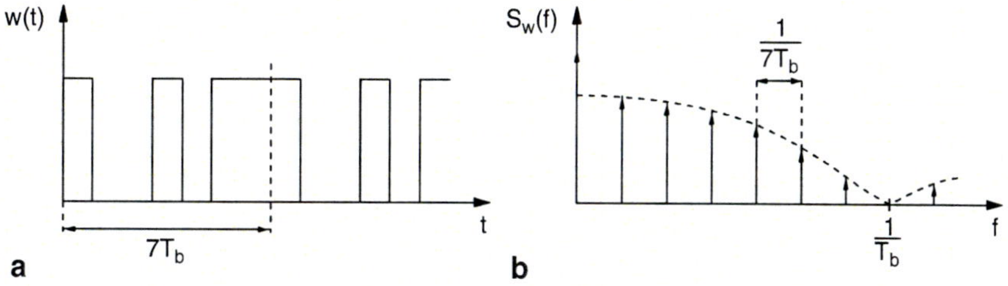

Fig. 2.5 PRBS as generated by the topology of Fig. 2.4a. **a** Waveform. **b** Spectrum

length is $2^3 - 1 = 7$. It can also be noticed that the maximum run length[1] equals 3. The number of ONEs and ZEROs differs by only one, so the pattern is said to be DC-balanced. In general, for an LFSR with m memory elements, the length of the generated PRBS is $2^m - 1$, the maximum run length equals m and the pattern is DC-balanced. A time domain representation of X_1, as generated by the topology in Fig. 2.4a, is shown in Fig. 2.5a.

The spectrum of a PRBS is fundamentally different from that of a truly random signal. Due to its repetitive character it contains only impulses. For the PRBS of Fig. 2.5a, the repetition rate equals $1/7T_b$. In the frequency domain, this is translated into a series of impulses with a mutual spacing of $1/7T_b$ which is shown in Fig. 2.5b. The height of every impulse is determined by an envelope that is formed by the spectrum of a single bit with a bit period T_b [Raz03].

A PRBS is typically used to scramble the actual data bits that need to be transmitted. This is done to eliminate a long run of ONEs or ZEROs which would cause the CDR to lose track of the clock signal and by doing so impede its proper operation. Another reason why a data signal is scrambled is to make sure that the spectrum of the actual transmitted signal is more or less constant in time. As a result, it is less dependent upon the actual message that is transmitted. This is advantageous to prevent interference with adjacent channels. On the contrary, a PRBS is almost always used to simulate and measure optical communication circuits because it is difficult to generate completely random binary waveforms. The PRBS is then used as a fictitious data signal. Typical run lengths for this vary between $2^5 - 1$ and $2^{31} - 1$.

2.3 The Laser Diode

At the transmitter side of an optical communication link, the data is converted to the optical domain by means of a laser diode. In contrast to a *light-emitting diode* (LED), a laser diode can be modulated at much higher speeds. Moreover, a laser diode can be viewed as a monochromatic light source as its output light energy is heavily concentrated around one particular wavelength. Furthermore, it produces an extremely

[1] The maximum number of consecutive ONEs or ZEROs.

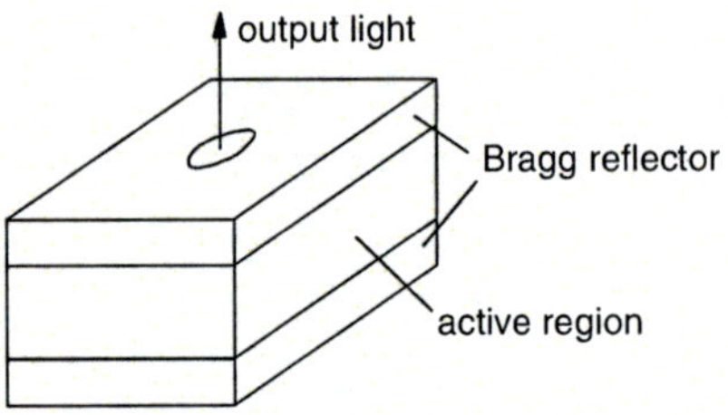

Fig. 2.6 Structural drawing of a VCSEL

focused and coherent[2] light beam that can travel a long distance before diverging. In the field of data communication, a *vertical-cavity surface-emitting laser* (VCSEL) is mostly preferred because of its low production cost. It is a semiconductor laser that emits its beam perpendicular to the top surface instead of from a side of the structure as in a conventional edge-emitting semiconductor laser. A structural drawing of a VCSEL is shown in Fig. 2.6. The light is generated inside the active region and is reflected back and forth between the *distributed Bragg reflectors* (DBR)[3]. Part of the generated light is able to escape through a hole in the upper reflector. The wavelength of the generated light of commercially available VCSELs is generally around 850 nm, which is short in comparison with other available types of lasers. They are mostly used for short communication links.

Electrically, a laser diode can be considered as a forward biased semiconductor diode. The relationship between the current through the laser, I_{laser}, and its optical output power is represented in Fig. 2.7. Up to the *threshold current*, I_{th}, the laser diode emits only incoherent light, just like a LED. If the current rises above this threshold, the lasing operation manifests itself. The optical output power P_{light} increases linearly with I_{laser}. The rate at which it rises is determined by $\delta P_{light}/\delta I_{laser}$, the *slope efficiency*. In contrast to a VCSEL, a LED has a zero threshold current and a lower slope efficiency.

Two techniques are available to generate a modulated optical signal: *external modulation* or *direct modulation*. When external modulation[4] is applied, the laser is actually a *continuous wave* (CW) laser that emits its light always at full power. The modulation process is performed by an external optical shutter that is open for a ONE and closed for a ZERO. A *Mach-Zehnder interferometer* is mostly used for this task. In a direct modulation scheme, the laser itself is turned on and off for a ONE and a ZERO respectively. Compared to external modulation, it is a simpler, more compact and cheaper modulation technique. For this reason, direct modulation is used throughout this work.

---

[2] Coherent light is light where all the photons have a definite phase relation to each other. This is due to the stimulated emission inside a laser.

[3] A DBR is a structure formed from multiple layers of alternating materials with different refractive indices. Its function is to reflect only light within a very narrow wavelength-band.

[4] External modulation is applied in systems that are less cost sensitive. It is able to output higher quality optical signals at higher bit rates.

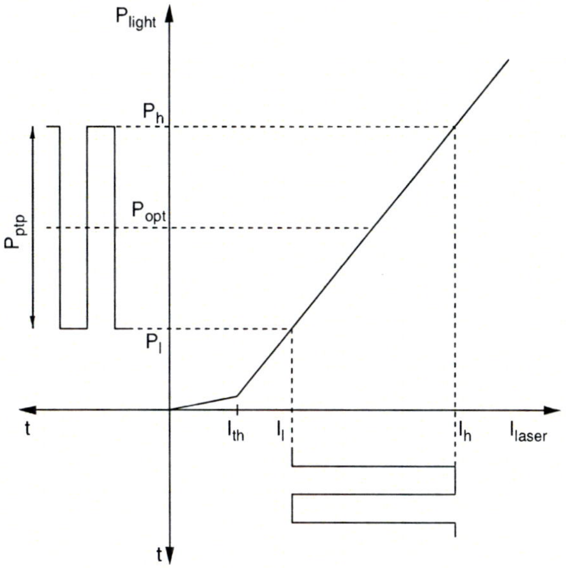

Fig. 2.7 Typical input-output characteristics of a VCSEL; the operation principle of direct modulation is also shown

In Fig. 2.7, the principle of direct modulation is shown. The current through the laser is modulated with the data and, as a consequence, the output light power varies between two levels: P_l and P_h for a ZERO and a ONE respectively. It can be noticed that the current through the laser is always larger than I_{th}. This is to minimize the turn-on delay t_{on} when switching from a ZERO to a ONE which can be approximated as follows [Säc05]:

$$t_{on} = \tau_c \ln \frac{I_h - I_l}{I_h - I_{th}}, \tag{2.6}$$

where τ_c represents the carrier lifetime which is typically around 3 ns. Equation (2.6) reveals that the turn-on delay decreases for increasing values of I_l. It drops to zero if I_l equals or exceeds I_{th}. A second reason why the laser should always be on is to minimize the jitter due to *relaxation oscillation*[5] [Raz03].

A faster operating speed is not the only consequence of choosing I_l above the threshold current of the laser diode. The other effect is a decreased optical signal swing for the same total transmitted power because the light is never turned off completely. This can be quantified by the *extinction ratio* r_e of the modulated laser diode:

$$r_e = \frac{P_h}{P_l}. \tag{2.7}$$

[5] Relaxation oscillation is an oscillation based upon the behavior of a system to return to equilibrium after being perturbed.

The average optical power for a ZERO and a ONE having an equal probability is given by:

$$P_{opt} = \frac{P_l + P_h}{2}. \tag{2.8}$$

This value is also shown in Fig. 2.7. **It is clarified in Sect.** 2.5.3 that the difference between P_h and P_l determines the quality of the received signal. This difference is called the peak-to-peak signal power P_{ptp}:

$$P_{ptp} = P_h - P_l. \tag{2.9}$$

Part of the average optical power P_{opt} is needed to bias the laser diode in its optimum operating region. The part that is effectively used to describe the signal is called the average signal power P_{sig}. For a ZERO and a ONE having the same probability, it is **defined as follows:**

$$P_{sig} = \frac{P_h - P_l}{2}. \tag{2.10}$$

The *power penalty* (*PP*) relates the average power to the average signal power:

$$P_{opt} = P_{sig} \cdot \underbrace{\frac{r_e + 1}{r_e - 1}}_{PP} \tag{2.11}$$

$$P_{sig}\,[\text{dB}] = P_{opt}\,[\text{dB}] - PP\,[\text{dB}]. \tag{2.12}$$

The relation between the extinction ratio and the power penalty is presented in Fig. 2.8. **The higher the extinction ratio, the more efficient the power is used because** a smaller portion of the total power is used to bias the laser diode.

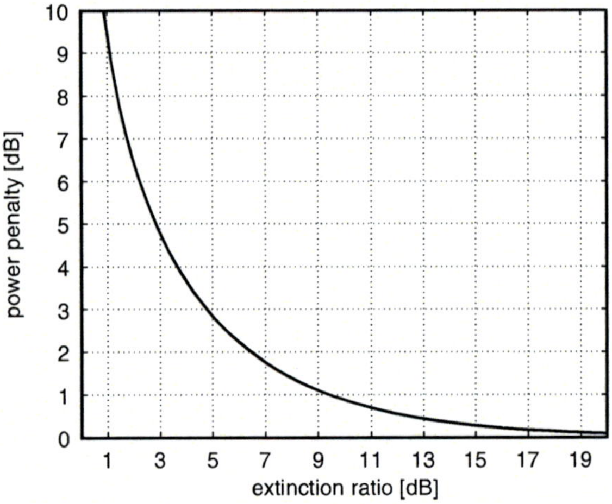

Fig. 2.8 The power penalty versus the extinction ratio

2.4 Optical Fiber

2.4.1 Silica Fiber

An optical fiber is used to guide the modulated light from the transmitter to the receiver. In Fig. 2.9, the typical structure of an optical fiber is shown. Three layers can be distinguished: *core*, *cladding* and *jacket*. The core guides the light while the cladding is needed to keep all the light inside the core. This principle is called *total internal reflection* and is fulfilled only when the angle of incidence is larger than a certain *critical angle*. The jacket protects the vulnerable structure of core and cladding. Usually, core and cladding are made from silica (SiO_2). To enable total internal reflection, the cladding should have a slightly lower *refractive index* than that of the core.

Two silica fiber types can be distinguished: *single-mode fiber* (SMF) and *multi-mode fiber* (MMF). The main structural difference between both is the diameter of the core which is between 8 and 10 μm for SMF and 50 or 62.5 μm for MMF. The cladding has a diameter of 125 μm in both fiber types. Due to the smaller core diameter of SMF, only one *mode*[6] can propagate through it. By contrast, several hundred modes can propagate in a MMF, each having a different path.

Loss A typical loss spectrum of silica fiber is plotted in Fig. 2.10. Two mechanisms can clearly be distinguished: *absorption* and *scattering*. Absorption is caused by the property of atoms and molecules to resonate at a specific frequency. In the infrared region[7] *ionic polarization* is the predominant effect, whereas in the ultraviolet region it is *electronic polarization* [Ina00]. The second type of losses, namely scattering losses, is caused by impurities in the core material. Scattered light waves leave the core if they do not fulfill the criterion of total internal reflection. This effect is called *Rayleigh scattering*. The scattering loss is proportional to

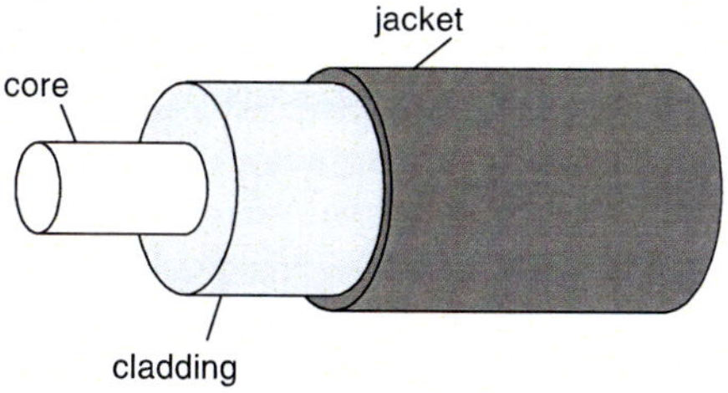

Fig. 2.9 Structure of an optical fiber

[6] There can be many modes traveling along a waveguide at a single frequency. Each mode has a different electric and magnetic field configuration. Ideally, there is no interaction between different modes [Olv92].

[7] Infrared radiation is electromagnetic radiation with a wavelength between 0.7 and 300 μm. Ultraviolet radiation is defined between 1 and 400 nm. Light with a wavelength in between the infrared and ultraviolet region is visible light.

Fig. 2.10 Typical loss spectrum of silica fiber

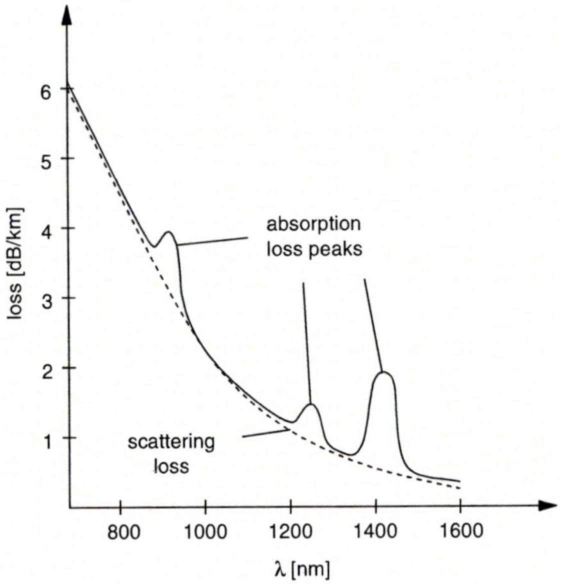

$1/\lambda^4$ [Raz03]. In order to encounter minimal loss, the wavelength should be approximately 1.5 μm. As will be discussed in Chap. 3, it is impossible to use a wavelength longer than 1 μm when the photodiode is integrated in silicon. Generally, because fiber loss increases linearly with the length of the fiber, it limits the distance between the transmitter and the receiver.

Dispersion Fiber dispersion also increases with the length of the fiber. However, it limits the distance between the transmitter and the receiver as well as the bit rate. There are three important dispersion mechanisms: *modal dispersion* (MD), *chromatic dispersion* (CHD) and *polarization-mode dispersion* (PMD).

The different paths in a MMF all have different path lengths. The direct path is the shortest way to the receiver and consequently arrives first, the wave that travels on the second shortest path arrives second and so on. This variation in the propagation delay between different modes causes the light pulses to spread out in the time domain which introduces *intersymbol interference* (ISI) and limits the bit rate. Due to the nature of a SMF that permits only a single mode in its core to propagate, this type of fiber is not subjected to MD.

CHD is a dispersion mechanism that affects both MMF and SMF. It is caused by a different propagation speed of light with different wavelengths (or colors). No CHD is present when light with a single wavelength is transmitted. However, every real laser diode emits a certain spectrum of wavelengths, though heavily concentrated in a very narrow band. On top of this, the modulation process of light with a single wavelength creates a spectrum in which a lot of different wavelengths are present. CHD is often quantified by the *dispersion parameter* D which is plotted in Fig. 2.11. Note that the minimal CHD is found at the band around a wavelength of 1300 nm.

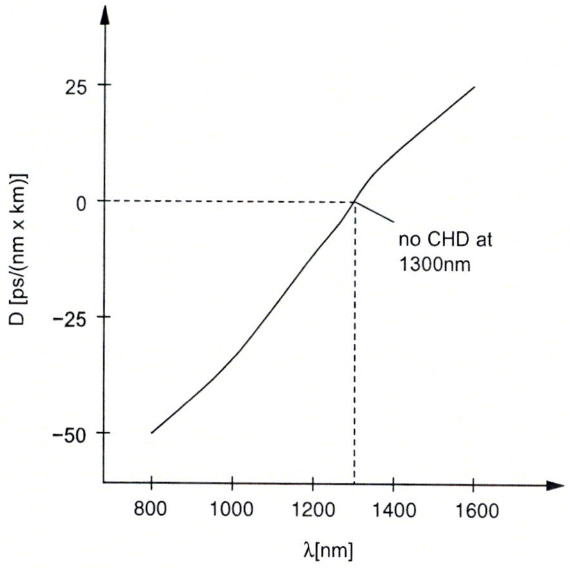

Fig. 2.11 Typical chromatic dispersion spectrum of silica fiber

To calculate the difference between the time of arrival of the fastest and the slowest wavelength for a transmitter with a spectral linewidth of $\Delta\lambda$ and for a fiber length L, the following formula can be used [Säc05]:

$$\Delta T = |D| \cdot \Delta\lambda \cdot L. \tag{2.13}$$

Because CHD is a linear phenomenon, its effect can be removed with the help of a *dispersion compensating fiber* (DCF).

Finally, PMD is caused by mechanical imperfections of the fiber such as a non-circular core or asymmetrical mechanical stress. This results in a difference in propagation speed between the horizontal and vertical polarization modes. Due to recent developments in fiber production, the delay between both modes can be kept below 1 ps for a fiber length of 100 km. Its effect can therefore be neglected in most cases.

2.4.2 Plastic Fiber

In contrast to the omnipresent silica fiber, *plastic optical fiber* (POF) is made out of plastic. It is the preferred choice in the growing market for consumer applications because an optical communication system making use of POF instead of silica fiber has a significantly lower cost thanks to the fiber, associated connectors and installation that are all inexpensive. This is a consequence of the large core diameter of POF, which is typically 1 mm. On top of this, POF has a high mechanical flexibility, which makes it an ideal solution to be used in cars (MOST[8]) or in houses (POF-PLUS[9]).

Compared to silica fiber, POF is characterized by a higher loss (up to 3 dB/m) which is caused by the less pure core material with relatively many impurities. Also, the bit rates that can be transported with POF are much lower than what can be obtained with a traditional fiber due to its higher dispersion. It is therefore predominantly applied in situations where the distance between transmitter and receiver is short and the needed bit rate is not extremely high.

Typically, the core material of POF is polymethyl methacrylate (PMMA) whereas fluoridated polymers are used in the cladding.

2.5 Optical Receiver Fundamentals

Due to fiber loss, the light signal at the receiver is greatly attenuated. Moreover, the signal is distorted owing to the dispersive character of the channel. The function of an optical receiver is to convert this attenuated and distorted light signal into digital data. The block diagram of such a receiver is shown in Fig. 2.12. The *photodiode* (Chap. 3) converts the received photons into an electrical current. In this work, the photodiode is integrated in the same silicon substrate that is also used to integrate the rest of the receiver chain. The *transimpedance amplifier* (TIA) (Chap. 4) amplifies the small photocurrent. The *equalizer* (Chap. 5) enhances the bandwidth of the receiver and more specifically that of the integrated photodiode. In the *post amplifier* (Chap. 6), the signal swing is enlarged to a rail-to-rail level. A *clock and*

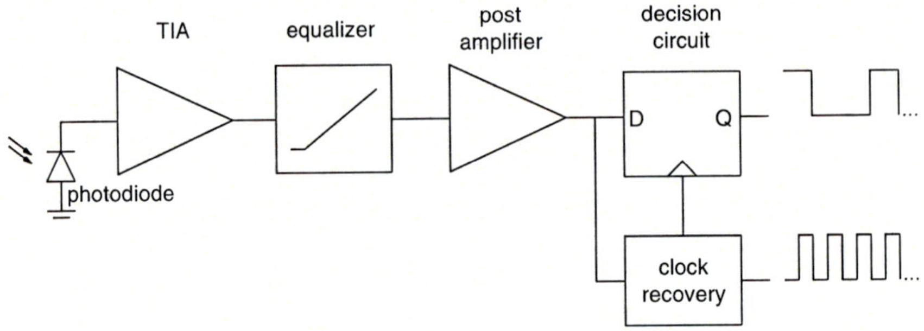

Fig. 2.12 Block diagram of an optical receiver; it provides the conversion of the attenuated modulated light signal at the receiver end of the optical fiber into a digital data stream with accompanying clock signal

[8] The media oriented systems transport (MOST) is a recent standard intended to connect multimedia devices with POF in cars.

[9] POF-PLUS is a European project that focuses on developing new photonic components and transmission technologies for large core POF systems, aiming at the unprecedented implementation of tens of Gbit/s transmission over this medium [POF10].

data recovery (CDR) system extracts the clock signal and retimes the data signal. The CDR is not treated in this work. Designing such a receiver implies making a lot of decisions. Therefore, a concise discussion about the most important trade-offs is given below.

2.5.1 Bandwidth Versus Bit Rate

The finite *bandwidth* of a system generally implies two effects: *low-pass filtering* and *high-pass filtering*. Whereas the former effect is always present in any real circuit implementation, the latter one can be absent, depending on the circuit topology. This is shown in Fig. 2.13 for the case where both effects are present. By definition, the bandwidth of a system is the frequency span within which the magnitude of the transfer function of the system is flat. The corner frequencies f_{hp} and f_{lp} are located where the magnitude is 3 dB lower than in the middle of the frequency band of interest. If $f_{hp} = 0$ Hz, the system is called a *baseband* system while it is called *passband* if $f_{hp} > 0$ Hz.

The receiver should be a baseband system if a truly random bit sequence is to be transmitted because this kind of signal has frequency components down to DC (Fig. 2.3b). On the contrary, when a PRBS is used as data pattern, the system can be passband if the corner frequency of the high-pass filtering is below the lowest frequency component in the PRBS (Fig. 2.5b)[10]. For a PRBS with a bit interval T_b and a maximum run length m, the condition is as follows:

$$f_{hp} < \frac{1}{(2^m - 1)T_b}. \tag{2.14}$$

For a PRBS with a maximum run length of 7 at a bit rate of 1 Gbit/s ($T_b = 1$ ns), f_{hp} should be below 7.81 MHz. However, even if condition (2.14) is fulfilled, *baseline wander* can be noticed. This is shown in Fig. 2.14a. The output level immediately follows the input level up to A but after course of time, the output level starts

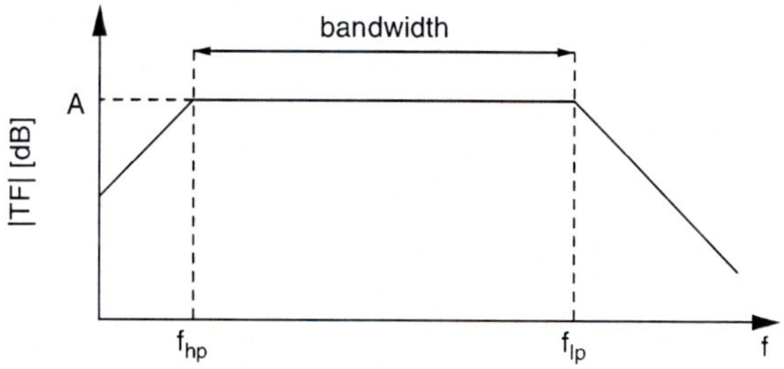

Fig. 2.13 Magnitude versus frequency of a system's transfer function on a logarithmic horizontal scale; the bandwidth is defined as the frequency span between f_{hp} and f_{lp}

[10] If the PRBS is not DC-balanced, an impulse at DC appears in the frequency domain (Fig. 2.5b). However, no information is contained in this impulse. Therefore, it can be removed without loss of signal integrity.

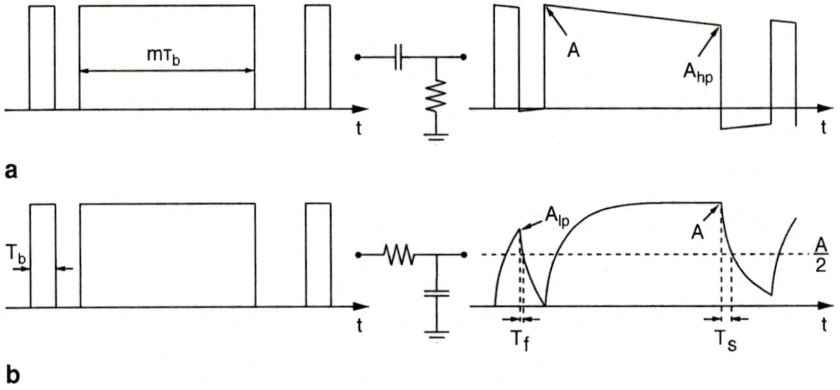

Fig. 2.14 Effect of a finite bandwidth on random digital data. **a** High-pass filtering. **b** Low-pass filtering

to droop significantly towards A_{hp}. The effect is most visible for the longest consecutive run of ONEs or ZEROs. If first-order high-pass filtering is assumed, the amount of baseline wander can be predicted as follows:

$$\frac{A_{hp}}{A} = e^{\frac{-mT_b}{\tau_{hp}}}, \qquad (2.15)$$

where τ_{hp} is the time constant of the equivalent first-order high-pass filter. In the example from above, suppose $f_{hp} = 7.81$ MHz ($\tau_{hp} = 20.4$ ns). It follows from (2.15) that $A_{hp} = 0.710 \cdot A$. The effect of high-pass filtering can also be seen as ISI because the output value of a particular bit is dependent upon its predecessors.

In contrast to high-pass filtering which only appears for certain topologies, low-pass filtering is always present in any real circuit implementation. As can be seen in Figs. 2.3b and 2.5b, the spectrum of a random or pseudo-random data pattern contains components up to an infinite frequency. It is important to notice that those components decrease in magnitude with increasing frequency. According to (2.5), the rate at which this happens is proportional to $1/f^2$. In Fig. 2.14b the effect of low-pass filtering on random digital data is shown. Because the actual response in a single bit interval depends on the previous bits, this effect can again be seen as ISI. In contrast to high-pass filtering, however, the effect is worst for a ZERO or ONE following a long run of ONEs or ZEROs, respectively. In this case, the output level is only able to rise up to A_{lp} whereas it would reach A if enough time were available. If first-order low-pass filtering is assumed, the relation between both levels can be quantified as follows:

$$\frac{A_{lp}}{A} = 1 - e^{\frac{-T_b}{\tau_{lp}}}, \qquad (2.16)$$

where τ_{lp} is the time constant of the equivalent first-order low-pass filter. For a PRBS at a bit rate of 1 Gbit/s ($T_b = 1$ ns) and a corner frequency $f_{lp} = 500$ MHz ($\tau_{lp} = 318$ ps), it follows from (2.16) that A_{lp} reaches A within 95.7%. Because the same effect manifests itself for a ONE to ZERO transition also, the actual vertical signal swing is only 91.4% of A.

A second effect of low-pass filtering, which can also be seen in Fig. 2.14b, is the variation of the threshold crossing dependent on the previous bits. This crossing can be fast, after T_f if a ZERO follows a single ONE or if a ONE follows a single ZERO. However, if a ZERO follows a long run of ONEs or if a ONE follows a long run of ZEROs, the crossing takes longer (T_s). This effect is known as *data-dependent jitter* (DDJ). The amount of DDJ can be determined with the following formula [Raz03]:

$$\frac{T_s - T_f}{T_b} = \frac{\tau_{lp}}{T_b} \cdot e^{\frac{-T_b}{\tau_{lp}}}. \tag{2.17}$$

The amount of DDJ can be calculated to be 1.38% of the bit interval for a PRBS at 1 Gbit/s ($T_b = 1$ ns) and a corner frequency $f_{lp} = 500$ MHz. To conclude, low-pass filtering affects the vertical (amplitude) scale as well as the horizontal (time) scale.

From now on, the term *bandwidth* is used as an alternative appellation for f_{lp} because in most practical implementations $f_{hp} \ll f_{lp}$.

2.5.2 *Noise Versus Bandwidth*

The objective of every receiver is to be able to process very small input signals. Therefore, the noise it generates should be sufficiently low. It is shown in Sect. 2.5.3 that a direct relation exists between the amount of noise generated by the receiver and the *bit error ratio* (BER) at the output of it. The quality of the output signal is thus not only affected by the finite bandwidth of the system (Sect. 2.5.2), but also by the amount of noise generated by it.

Every resistor, diode and transistor generates noise. The noise of a resistor is especially *white noise* which is generated by thermal agitation of charge carriers inside the resistor material. While white noise is dominant for a resistor, some *pink noise* is also present. The voltage noise spectral density of a resistor with resistance R is given by [San06]:

$$\overline{dV_{n,R}^2} = \underbrace{4kTR \cdot df}_{\text{white noise}} + \underbrace{V_R^2 \frac{K_R R_\square}{A_R} \cdot \frac{df}{f}}_{\text{pink noise}}, \tag{2.18}$$

where k is the *Boltzmann's constant*[11], T is the temperature of the resistor, V_R is the DC voltage over the resistor, K_R is a fitted parameter for a specific kind of resistor, $R_\square$ is the resistance per square[12] and A_R is the area of the resistor.

[11] Boltzmann's constant ($k = 1.38 \cdot 10^{-23}$ J/K) is a physical constant relating energy at the particle level with temperature at the bulk level.

[12] In integrated circuit technology, resistors are often described in terms of their resistance per square which is the resistance of a square of the resistor material. The resistance per square is not dependent on the size of the square. This is a convenient parameter in integrated circuits because it is essentially a two-dimensional environment.

A diode also generates both types of noise components. The white noise is in this case *shot noise* which is not dependent on temperature. The pink noise is again determined by the area of the device. The total current noise spectral density of a diode that conducts a current I_D is given by:

$$\overline{dI_{n,D}^2} = 2qI_D \cdot df + I_D \frac{K_D}{A_D} \cdot \frac{df}{f}, \tag{2.19}$$

where q is the elementary charge, K_D is a fitted parameter for the specific kind of diode and A_D is the area of the diode. The white and pink noise can again be distinguished clearly. In addition to this current noise, a diode also has a series resistance with its accompanying noise source. However, this noise contribution is not considered in the remainder of this book.

The predominant white noise component of a MOSFET is channel noise. Because the channel is resistive, the noise mechanism is thermal and consequently white. Other white (thermal) noise components in a MOSFET are generated by the gate resistance R_g, the source resistance R_s and the substrate resistance R_b. Because the current in a MOSFET flows close to the surface where a lot of defects in the crystal structure are present, a MOSFET also generates a considerable amount of pink noise. The gate referred noise voltage spectral density of a MOSFET is given by:

$$\overline{dV_{n,T}^2} = 4kTR_{eff} \cdot df + \frac{K_T}{WLC_{ox}^2} \cdot \frac{df}{f}, \tag{2.20}$$

where K_T represents a fitted parameter for a specific transistor type in a specific process technology, W and L are the gate width and length respectively and C_{ox} is the gate oxide capacitance per unit area. R_{eff} is the equivalent resistance that generates the same amount of white noise as the channel, gate, source and bulk together. Because all these noise sources are not correlated, the effective resistance can be calculated as follows [San06]:

$$R_{eff} = \frac{\gamma}{g_m} + R_g + R_s + R_b(n-1)^2, \tag{2.21}$$

where g_m is the transconductance of the transistor and the parameter n has a value between 1.2 and 1.5. The terms corresponding to the gate, source and bulk resistance can be distinguished clearly. The first term corresponds to the channel which has an equivalent noise resistance of γ/g_m. The parameter γ equals 2/3 for long-channel transistors but increases due to velocity saturation in short-channel devices. From now on, the parameter γ is supposed to be equal to 4/3, a value which is valid for a 130 nm CMOS technology [San06].

A circuit consists of a number of noise generating components which all contribute to the total output noise. To calculate this output noise, the transfer function of every noise source to the output needs to be determined. However, the total output noise voltage spectral density of a circuit depends on the configuration of the input as shown in Fig. 2.15. If the input is left open or is short-circuited, the respective output noise

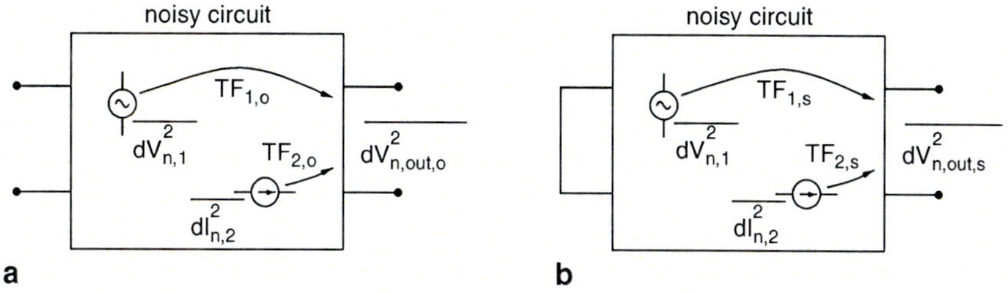

Fig. 2.15 The output noise of a circuit consisting of several internal noise sources. **a** The output noise if the input is left open. **b** The output noise when the input is short-circuited

voltage spectral densities $\overline{dV_{n,out,o}^2}$ and $\overline{dV_{n,out,s}^2}$ can be calculated with the following formulas:

$$\overline{dV_{n,out,o}^2} = \overline{dV_{n,1}^2} \cdot |TF_{1,o}|^2 + \overline{dI_{n,2}^2} \cdot |TF_{2,o}|^2 + \cdots, \tag{2.22}$$

$$\overline{dV_{n,out,s}^2} = \overline{dV_{n,1}^2} \cdot |TF_{1,s}|^2 + \overline{dI_{n,2}^2} \cdot |TF_{2,s}|^2 + \cdots, \tag{2.23}$$

where $TF_{n,o}$ and $TF_{n,s}$ denote the transfer function of the nth noise source to the output when the input is left open or is short-circuited respectively. Another way to represent the circuit noise is by its equivalent input noise. In this case, the circuit is assumed to be completely noiseless and the real output noise is assumed to be generated by an external voltage and current source as shown in Fig. 2.16 for a circuit where the input is connected to a source impedance Z_s. It can be proven that this can be done for every linear circuit [Raz01]. The equivalent input noise voltage and current spectral densities are calculated as follows:

$$\overline{dV_{n,in}^2} = \frac{\overline{dV_{n,out,s}^2}}{|TF_V|^2}, \tag{2.24}$$

$$\overline{dI_{n,in}^2} = \frac{\overline{dV_{n,out,o}^2}}{|TF_I|^2}, \tag{2.25}$$

where TF_V and TF_I represent the transfer function for an input voltage signal and for an input current signal respectively. Obviously, the input noise current should be

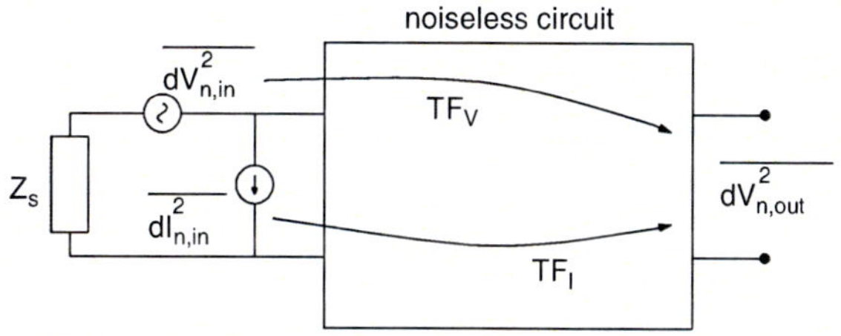

Fig. 2.16 The equivalent input noise of a circuit when the input is connected to a source impedance Z_s

calculated by taking into account $\overline{dV_{n,out,s}^2}$ alone because the voltage source at the input has no effect if the input is left open as in Fig. 2.15a. Equivalently, the input current noise is calculated by taking into account $\overline{dV_{n,out,o}^2}$ alone because the current source at the input has no effect if the input is short-circuited as in Fig. 2.15b. If the magnitude of Z_s in Fig. 2.16 has an infinite value, the voltage noise source $\overline{dV_{n,in}^2}$ has no effect while the noise current $\overline{dI_{n,in}^2}$ flows completely through the input nodes of the circuit, generating by doing so the total output noise. On the contrary, the current noise source is short-circuited if the magnitude of Z_s is zero. For a finite magnitude of Z_s, both the voltage and the current source generate part of the total output noise.

The noise calculations from above determine only the spectral densities of the noise. To obtain the *root mean square* (RMS) value of the output noise, the output noise spectral density of Fig. 2.16 needs to be integrated over the entire frequency spectrum:

$$V_{n,RMS} = \sqrt{\int_0^\infty \overline{dV_{n,out}^2}}. \tag{2.26}$$

This operation is visualized in Fig. 2.17 where the shaded area represents the squared integrated output noise voltage. Due to the physical limitations of every circuit implementation, every noise spectrum eventually drops to zero. Therefore, the integral of (2.26) always leads to a finite value[13].

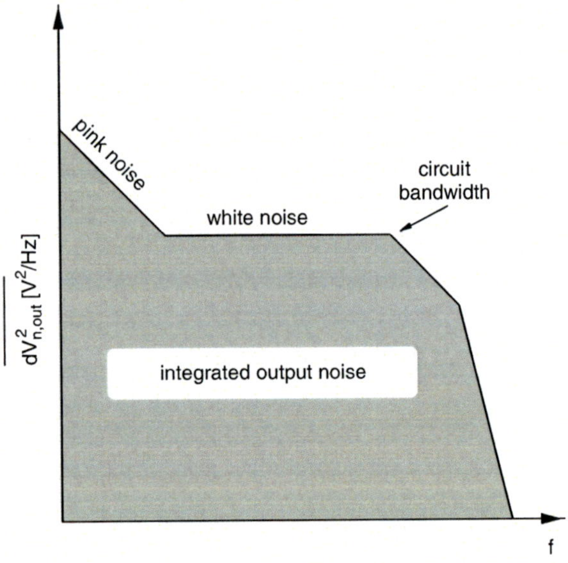

Fig. 2.17 Integration of the output noise voltage spectral density gives the total squared output noise voltage. Note that both the horizontal and the vertical axes have a logarithmic scale

[13] In the case of pink noise, the integral of (2.26) results in an infinitely high RMS value. Consequently, noise can only be pink over a limited range of frequencies, excluding 0 Hz.

It can be derived from (2.26) that the output noise voltage $V_{n,RMS}$ increases if the circuit bandwidth increases. To illustrate this, assume that a circuit contains only one resistor with resistance R_{noise} that generates only white noise with a noise spectral density of $\overline{dV_{n,R_{noise}}^2} = 4kTR_{noise}$. Assume also that the transfer function to the output for the noise of this resistor is a first-order low-pass filter with a time constant $\tau_{circuit}$. The RMS noise voltage at the output can be calculated as follows:

$$\begin{aligned} V_{n,RMS} &= \sqrt{\int_0^{\infty} \overline{dV_{n,R_{noise}}^2} \cdot \left| \frac{1}{1 + j2\pi f \tau_{circuit}} \right|^2} \\ &= \sqrt{4kTR_{noise} \int_0^{\infty} \frac{1}{1 + (2\pi f \tau_{circuit})^2}\, df} \\ &= \sqrt{\frac{kTR_{noise}}{\tau_{circuit}}}. \end{aligned} \tag{2.27}$$

If $R_{noise} = 50\ \Omega$ and $f_{circuit} = 1$ GHz ($\tau_{circuit} = 159$ ps), (2.27) results in an RMS output noise voltage of 36.1 μV_{RMS} at room temperature (300 K). If the bandwidth is increased to 2 GHz, the RMS output noise voltage rises with a factor of $\sqrt{2}$ to 51.0 μV_{RMS}.

In Sect. 2.5.1 it has been shown that the bandwidth should be as high as possible to minimize the degradation of the signal quality in terms of signal swing and DDJ. However, in this section it is found that a higher circuit bandwidth results in a larger RMS noise voltage, which in its turn degrades the quality of the output signal. While a high bandwidth seems advantageous if the input signal swing is large so that noise is of subordinate importance, a smaller bandwidth is better to cope with very weak input signals that are easily affected by circuit noise. In conclusion, the objective in choosing an appropriate receiver bandwidth is to make it as small as possible while still retaining enough vertical signal swing and repressing DDJ. In Sect. 2.5.1 it has been concluded that for a bit rate of 1 Gbit/s and a bandwidth of 500 MHz, the DDJ consumes only 1.38% of the bit interval and the vertical signal swing is 91.4% of that of a receiver with an infinite bandwidth. Therefore, a bandwidth of B/2 Hz seems sufficient not to introduce too much signal distortion for a data stream at a bit rate of B bit/s. In an actual implementation, the receiver is very likely to have additional poles above the circuit bandwidth. These poles increase the DDJ and reduce the signal swing further. A bandwidth of 0.6–0.7 · B Hz is therefore assumed to be optimal to receive data at a bit rate of B bit/s [Säc05].

2.5.3 *Bit Error Ratio Versus Noise*

As discussed in Sect. 2.5.2, the RMS noise voltage depends on the bandwidth of the system. However, it is not clear yet what the effect of this noise voltage is on the BER. To examine this, the circuit is assumed to have a bandwidth which is high enough not to distort the shape of the signal. Consequently, the signal transitions are performed infinitely fast, as visualized in Fig. 2.18. In the absence of noise, only two discrete values are possible for the signal amplitude: V_h for a ONE and V_l for a ZERO. This is shown in Fig. 2.18a where the data and the conditional *probability density function* (PDF) of its amplitude are plotted. When a ZERO or a ONE is received, the amplitude is always V_l or V_h respectively. The conditional chance of this event is therefore 1 in both cases. In the receiver, the decision whether a ZERO or a ONE is received is based upon the following condition:

$$\text{if } V \leq V_t \Rightarrow \text{ZERO} \tag{2.28}$$

$$\text{if } V > V_t \Rightarrow \text{ONE.} \tag{2.29}$$

V_t is called the decision threshold voltage. In the absence of noise, there is no chance of making a wrong decision (2.28)–(2.29) if the decision threshold V_t is located anywhere between V_h and V_l. As a result, the corresponding BER is 0.

Things change quite drastically when noise is taken into account. This situation is visualized in Fig. 2.18b. The signal transitions still happen infinitely fast, so noise has no influence on the decision threshold crossing time. Between the signal transitions, the amplitude is now not well-defined any more because the noise voltage is added to the signal. To simplify the analysis, only white noise is considered here. The PDF

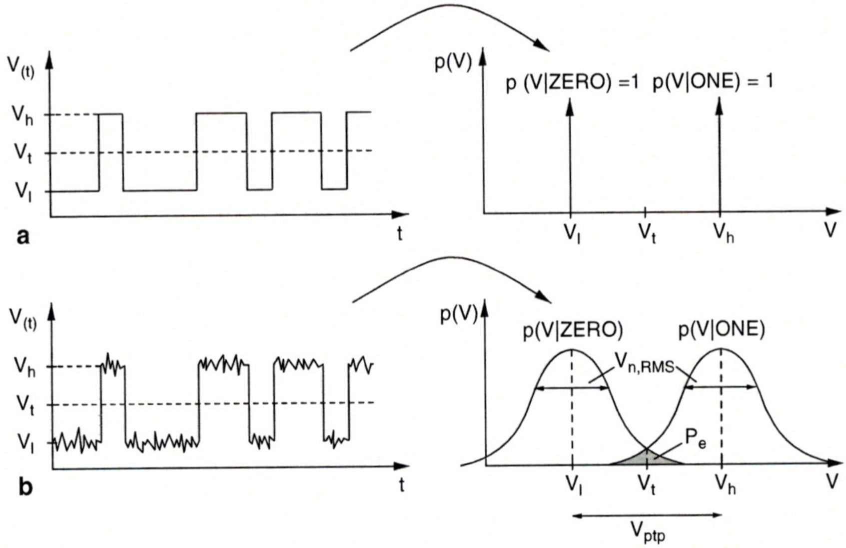

Fig. 2.18 The effect of noise on a random data pattern. **a** Data pattern and its corresponding PDF when no noise is present; the receiver always makes the correct decision if V_t is between V_h and V_l. **b** Data pattern and its corresponding PDF when noise is present; the receiver makes an errors with a probability P_e

of white noise is characterized by a *Gauss function* with an average of 0 and $V_{n,RMS}^2$ as variance:

$$p(V_n) = \frac{1}{2\pi V_{n,RMS}} e^{-V_n^2/2V_{n,RMS}^2}. \tag{2.30}$$

The simple PDF of the signal amplitude when noise is not taken into account Fig. 2.18a is now transformed into the PDF in Fig. 2.18b. When a ZERO is received, the PDF of the signal amplitude is $p(V|\text{ZERO})$ and $p(V|\text{ONE})$ when a ONE is received. These conditional distributions can be calculated respectively by means of the following formulas:

$$p(V|\text{ZERO}) = \frac{1}{2\pi V_{n,RMS}} e^{-(V_n - V_l)^2/2V_{n,RMS}^2} \tag{2.31}$$

$$p(V|\text{ONE}) = \frac{1}{2\pi V_{n,RMS}} e^{-(V_n - V_h)^2/2V_{n,RMS}^2}. \tag{2.32}$$

For the remainder of the derivation, it is assumed that the noise voltage $V_{n,RMS}$ does not depend on whether a ZERO or a ONE is received, which is a reasonable assumption in most cases[14]. In contrast to the situation without noise, there is always a possibility of making a wrong decision because the conditional PDFs only decrease asymptotically to 0. $P_{e,\text{ZERO}}$ is the probability that the receiver decides that a ONE has been transmitted while actually a ZERO is transmitted. $P_{e,\text{ONE}}$ is the probability that the receiver decides that a ZERO has been transmitted while a ONE is transmitted. Both probabilities can be expressed mathematically as follows:

$$P_{e,\text{ZERO}} = \int_{V_t}^{\infty} p(V|\text{ZERO})\, dV \tag{2.33}$$

$$P_{e,\text{ONE}} = \int_{-\infty}^{V_t} p(V|\text{ONE})\, dV. \tag{2.34}$$

For a ZERO and a ONE having an equal probability to be transmitted, the combined chance on a bit error is as follows:

$$P_e = \frac{1}{2} P_{e,\text{ZERO}} + \frac{1}{2} P_{e,\text{ONE}}. \tag{2.35}$$

This bit error probability is usually called the BER. It is also shown in Fig. 2.18b as the shaded area below the two conditional PDFs. To minimize P_e, the decision threshold must be chosen very carefully. If a ZERO and a ONE occur with equal

---

[14] When the receiver noise is dominated by the shot noise of the photodiode, the noise voltage for a ONE is larger than that of a ZERO because the shot noise is dependent on the current through the photodiode. This effect is neglected here.

probability and if the RMS noise voltage is equal for both logical levels, the decision threshold should be chosen exactly in the middle of V_l and V_h:

$$V_t = \frac{V_h + V_l}{2}. \tag{2.36}$$

With the help of these formulas, the BER can be written as [Cou02]:

$$P_e = Q\left(\frac{V_{ptp}}{2V_{n,RMS}}\right), \tag{2.37}$$

where V_{ptp} is the peak-to-peak signal amplitude $V_h - V_l$ (Fig. 2.18b). $Q(x)$ is called the *Q-function* and is the cumulative distribution function for the Gaussian distribution. The argument of the Q-function in (2.37) represents the *signal-to-noise-ratio* (SNR) of the signal. The graphical representation of (2.37) is given in Fig. 2.19 for a range of SNRs. As a rule of thumb, the BER decreases by two orders of magnitude for each decibel increase in SNR [Raz03]. For example, if the required BER is 10^{-12} or lower, the argument of the Q-function in (2.37) needs to be equal to or higher than 7. If the noise of the signal has an RMS value of 1 mV_{RMS}, the required peak-to-peak signal voltage needs to be larger than 14 mV (2.37)[15].

To quantify the BER as a function of the SNR (2.37), it is assumed that the bandwidth of the system is high enough in order not to introduce any signal distortion. However, as mentioned in Sect. 2.5.2, the bandwidth should be 0.6–0.7 · B Hz for a bit rate of B bit/s as an optimal trade-off between signal distortion and noise. As derived in Sect. 2.5.1, **a finite bandwidth has two important consequences: DDJ and**

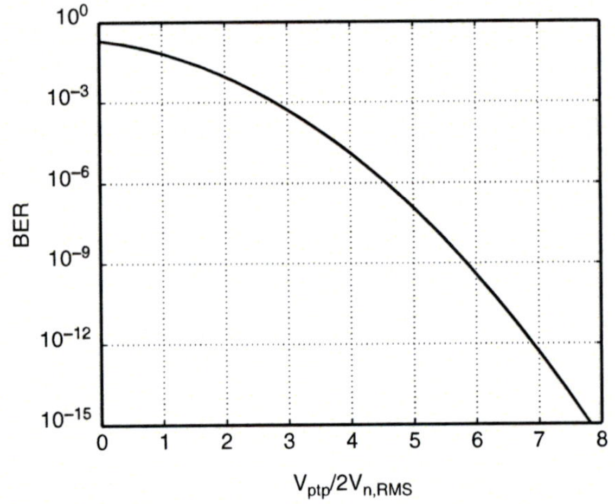

Fig. 2.19 Representation of the Q-function; with $V_{ptp}/2V_{n,RMS}$ as its argument the Q-function determines the BER of a signal corrupted by white noise

[15] The BER at the output of a receiver is determined by the signal quality at the input of this receiver as well as by the extra noise that is added by the receiver. In the remainder of this book, it is assumed that the former can always be neglected compared to the receiver noise.

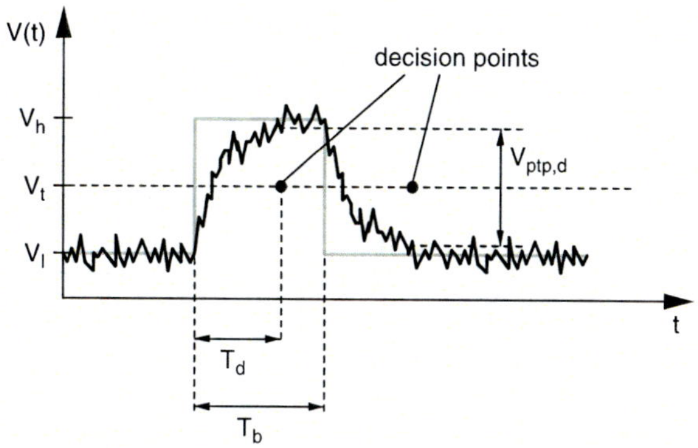

Fig. 2.20 The effect of noise and a limited bandwidth on a random data signal; the effective SNR decreases due to the finite bandwidth

a reduced signal swing. Some remarks are therefore needed to put things in a correct perspective. The effect of noise and a finite bandwidth on a random data pattern is shown in Fig. 2.20.

Without distortion, it is not important where exactly in the bit interval the decision is made as to whether a ZERO or a ONE has been transmitted. If the bandwidth is limited, this is not the case any more. DDJ and the reduced signal swing cause the optimal sampling point to be somewhere in the middle of the bit interval. This is especially important considering the fact that noise now also influences the rising and falling edges due to their finite rise and fall time. In addition to DDJ, an extra jitter component appears that is, in contrast to DDJ, random in nature because it is generated by noise. Therefore, it is called *random jitter* (RJ). A second effect of the limited bandwidth is a decreased vertical signal swing and from that a decreased SNR when compared to a signal with an infinite bandwidth. To be correct, the calculation of the BER should be done with the actual SNR. To quantify this, the vertical signal swing at the decision point, $V_{ptp,d}$, should be known. Its value is represented in Fig. 2.20. Obviously, $V_{ptp,d}$ is always smaller than V_{ptp}, the signal swing for a system with an infinite bandwidth. A more precise calculation of the BER can be obtained by altering (2.37) into the following formula:

$$P_e = Q\left(\frac{V_{ptp,d}}{2V_{n,RMS}}\right). \tag{2.38}$$

As described by (2.16), the vertical signal swing at the end of the bit interval is reduced because of a limited bandwidth. However, to calculate the BER, the signal swing at the decision point is important. Therefore, (2.16) is transformed into the following formula to calculate the effective signal swing at the decision point:

$$\frac{V_{ptp,d}}{V_{ptp}} = 1 - 2e^{\frac{-T_d}{\tau_{lp}}}. \tag{2.39}$$

A factor 2 is added because the swing is limited for a ZERO to ONE as well as for a ONE to ZERO transition. For a PRBS at a bit rate of 1 Gbit/s (T_b = 1 ns) and a corner frequency (bandwidth) $f_{lp} = 500$ MHz ($\tau_{lp} = 318$ ps), it follows from (2.39) that $V_{ptp,d} = 0.914 \cdot V_{ptp}$ if the decision point is at the end of the bit interval ($T_d = T_b$). To obtain a BER below 10^{-12}, it has been derived before that a peak-to-peak signal swing of 14 mV is sufficient if the RMS noise voltage equals 1 mV$_{RMS}$. However, the BER increases to $3.14 \cdot 10^{-11}$ (2.38–2.39) when the finite bandwidth is taken into account. If noise generates too much RJ, it is necessary to sample before the end of the bit interval. If, for example, the decision point is moved forward so that $T_d = 0.75 \cdot T_b$, it follows from (2.39) that $V_{ptp,d} = 0.811 \cdot V_{ptp}$. The BER then equals $2.85 \cdot 10^{-9}$. Clearly, both the integrated noise as well as the bandwidth determine the BER of the receiver.

Sensitivity Another way to quantify the performance of an optical receiver can be obtained by considering its *sensitivity* P_s. This is defined as the minimal average optical input power that is needed to achieve a certain BER. To determine the sensitivity, the conversion from the light signal to the electrical output signal is described as follows:

$$V_{ptp} = P_{ptp} \cdot R_{pd} \cdot Z_r, \tag{2.40}$$

where R_{pd}[A/W] is the *responsivity* of the photodiode and $Z_r[\Omega]$ represents the transimpedance gain of the receiver. The same can be done for the RMS noise:

$$V_{n,RMS} = P_{n,RMS} \cdot R_{pd} \cdot Z_r. \tag{2.41}$$

$P_{n,RMS}$ can be considered as the equivalent integrated input noise of the receiver. The BER can now be calculated also at the input by transforming (2.37) into:

$$P_e = Q\left(\frac{P_{ptp}}{2P_{n,RMS}}\right). \tag{2.42}$$

If the sensitivity is defined at a BER of 10^{-12} for which the argument of the Q-function needs to be 7 or higher, P_s equals the peak-to-peak signal swing that is needed to obtain this: $P_s = 7 \cdot 2P_{n,RMS}$. Taking into account the finite extinction ratio of the laser diode, the total optical input power to achieve a certain BER can be calculated as follows:

$$P_{opt} = P_s + PP, \tag{2.43}$$

where PP is the power penalty due to the finite extinction ratio of the laser diode.

2.6 Conclusion

An optical communication system is used to transmit random digital data. At the transmitter, a laser diode converts the signal from the electrical domain to the optical domain. Due to the fact that the current through a laser diode should always be larger than a certain threshold current, the average transmitted power always exceeds the average signal power. The optical signal is guided to the receiver by means of an optical fiber. Mostly, a silica fiber is used because of its low loss and dispersion. However, in some specific application domains plastic fiber is applied which is characterized by a relatively high loss and dispersion but, on the contrary, also by a very low cost price.

The optical receiver, which is the topic of this thesis, is charged with the conversion of the attenuated and distorted optical signal into a truly digital bit stream that can be interpreted by the digital back-end. One of the key parameters of an optical receiver is its bandwidth which needs to be high enough to prevent the signal from being distorted. This distortion has two important consequences: a reduced signal swing and the appearance of DDJ. On the contrary, the bandwidth should also be as low as possible to guarantee low noise operation. Eventually, the most important parameter of an optical receiver is its BER. This probability on a bit error is a strong function of the ratio of the peak-to-peak signal swing and the RMS noise. Both the peak-to-peak signal swing and the integrated noise depend on the bandwidth of the receiver. An optimal trade-off between a good low-noise performance and a low distortion is obtained when, for a bit rate of B bit/s, the bandwidth equals $0.6\text{–}0.7 \cdot B$ Hz. In conclusion, the BER performance combines the effect of a finite bandwidth and the RMS noise. The sensitivity determines the required optical input power to obtain a certain BER.

Chapter 3
From Light to Electric Current—The Photodiode

The first task of an optical receiver is to convert the signal from the optical domain to the electrical domain. This is performed by the photodiode which is discussed thoroughly in this chapter.

First, the principle of carrier generation and recombination in a semiconductor is explained. Thereafter, the concept of a photodiode is described in order to collect the carriers that are generated by impinging light as a photocurrent.

Next, the discussion is concentrated on CMOS photodiodes which have serious issues concerning their responsivity and especially their speed. Several techniques are proposed to alleviate these problems. Finally, it is shown that the dielectric layers on top of the silicon substrate in a CMOS process can have a detrimental effect on the performance of an integrated photodiode.

3.1 Working Principle

3.1.1 Carrier Generation and Recombination

Every semiconductor, whether it is intrinsic or doped, has an equilibrium state in which the generation rate of new electrons and holes equals the recombination rate of both. In this situation, the concentration of electrons n and the concentration of holes p is subjected to the following condition:

$$n \cdot p = n_i^2, \tag{3.1}$$

where n_i stands for the intrinsic carrier concentration. This intrinsic carrier concentration is a material constant which depends only on temperature. For silicon at room temperature n_i equals roughly 10^{10} cm^{-3}. In an intrinsic semiconductor, the concentrations of electrons and holes are equal to each other and therefore are also equal to the intrinsic carrier concentration (3.1). By contrast, the concentration of electrons or holes is artificially increased in a doped semiconductor. It follows from (3.1) that the concentration of holes or electrons decreases accordingly. For example, if a piece

F. Tavernier, M. Steyaert, *High-Speed Optical Receivers with Integrated Photodiode in Nanoscale CMOS,* Analog Circuits and Signal Processing 5,
DOI 10.1007/978-1-4419-9925-2_3,

of silicon has a concentration of free electrons of 10^{17} cm^{-3}, the concentration of free holes is only 10^3 cm^{-3}. The electrons are then called *majority carriers* while the holes are called *minority carriers*. The material itself is described as *n-type* silicon.

At equilibrium, the continuous thermal generation of electron-hole pairs is compensated by the continuous recombination of electron hole pairs so that (3.1) is always fulfilled. Any deviation from this condition causes an increased net generation or recombination in order to re-establish the equilibrium state. In addition to thermal generation, which is an internal mechanism, electron-hole pairs can be created by external mechanisms of which the most common is *radiative generation*. This mechanism is applied usefully to build photodetectors, as will be shown later. The counteracting force to this is provided by a recombination mechanism with *Shockley-Read-Hall recombination* being the most important one in silicon.

Radiative Generation Electromagnetic radiation can generate electron-hole pairs in a semiconductor. Equivalently, the recombination of an electron and a hole can create electromagnetic radiation. The first mechanism is radiative generation, while the second one is radiative recombination. Both processes are represented in Fig. 3.1. Radiative generation causes the concentration of electrons and holes to rise while radiative recombination reduces both concentrations.

An electron-hole pair is created in the semiconductor crystal when an electron is moved from the valence band into the conduction band. In this way, a bounded electron is transformed into a free electron which contributes to the conductivity of the semiconductor. In addition to this, a hole, which is an electron vacancy, is created in the valence band thereby increasing the semiconductor conductivity. To generate an electron-hole pair, the energy level of the electron needs to rise by an amount larger than or equal to the energy difference between the valence band and the conduction band, namely the band gap energy E_g. An impinging electromagnetic wave can provide this energy on condition that it has a photon energy $h\nu$ which is at least as high as the band gap energy of the semiconductor under consideration. This

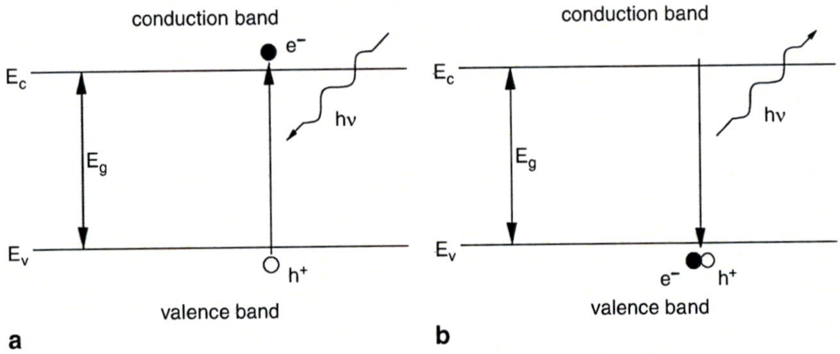

Fig. 3.1 Radiative generation and radiative recombination mechanism in a semiconductor with a band gap energy E_g. **a** Radiative generation. **b** Radiative recombination

condition can be expressed mathematically as follows:

$$h\nu \geq E_g = E_c - E_v, \tag{3.2}$$

where h is Planck's constant[1], ν is the frequency of the electromagnetic wave, E_c is the lowest energy level of the conduction band, and E_v is the highest energy level of the valence band. From condition (3.2), it follows that a photon only needs to possess an energy level that is as high as the band gap energy. The amount of excess energy per photon is dissipated as heat. To maximize the number of electron-hole pairs created by an impinging electromagnetic wave with a certain power, the frequency of that wave should be chosen as follows:

$$\nu = \frac{E_g}{h}. \tag{3.3}$$

The band gap energy of silicon at room temperature is 1.12 eV which means that an impinging electromagnetic wave should have a frequency of at least 271 THz in order to create electron-hole pairs. The frequency ν and wavelength λ of an electromagnetic wave are related to each other by the following formula:

$$c = \lambda \cdot \nu, \tag{3.4}$$

where c is the propagation speed of the wave. In a vacuum, this speed is constant and is equal to $300 \cdot 10^6$ m/s. With (3.4), the wavelength corresponding to a frequency of 271 THz can thus be calculated to be 1.11 μm, which is in the near-infrared region of the electromagnetic spectrum. Summarizing, an electromagnetic wave with a wavelength below 1.11 μm in vacuum can create an electron-hole pair in silicon at room temperature.

While penetrating into the semiconductor material, the electromagnetic wave is gradually absorbed. The *absorption coefficient* α determines the fraction of photons that is absorbed, and thus transformed into electron-hole pairs, per unit length. This mechanism is mathematically characterized by *Lambert-Beer's Law*:

$$\Phi(d) = \Phi_0 \cdot e^{-\alpha d}, \tag{3.5}$$

where $\Phi(d)$ is the photon flux at a depth d in the semiconductor and Φ_0 is the photon flux at the semiconductor surface. It can be derived from (3.5) that, after penetrating the semiconductor over a distance of $1/\alpha$, the remaining photon flux is only $e^{-1} = 36.8\%$ of the initial photon flux. The distance $1/\alpha$ is called the *penetration depth*. The absorption coefficient is a material constant which depends strongly on wavelength. This is clearly visible in Fig. 3.2, where the absorption coefficient of silicon is plotted as a function of wavelength. Photons with a wavelength above 1.11 μm have a very low absorption coefficient and, as a result, a very long penetration depth. This is due to their photon energy being insufficient to create electron-hole

[1] Planck's constant is a physical constant used to describe the sizes of quanta in quantum mechanics. Its value is $6.63 \cdot 10^{34}$ Js.

Fig. 3.2 The absorption coefficient of silicon as a function of wavelength [Inc10]

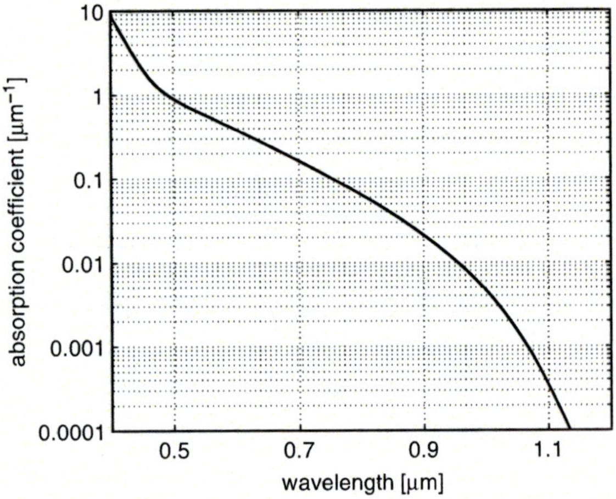

pairs. These photons can travel through the semiconductor material without being absorbed significantly by it. Photons with a wavelength below 1.11 μm have enough energy to make electrons bridge the band gap and be absorbed by the semiconductor. However, it is clear from Fig. 3.2 that this process evolves gradually: the absorption coefficient rises steadily for decreasing wavelengths. The normalized photon flux as a function of depth for silicon, as predicted by (3.5), is plotted in Fig. 3.3 for three different wavelengths[2]. It can be concluded from this figure that the photon wavelength has a huge impact on the photon flux profile. For a wavelength of 850 nm,

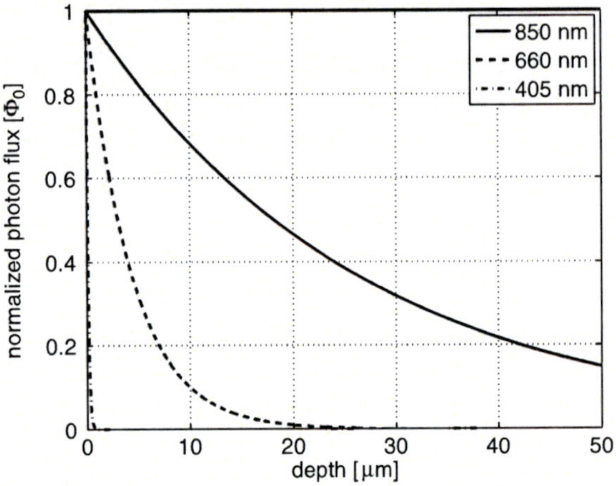

Fig. 3.3 The normalized photon flux as a function of depth for three different wavelengths. The penetration depth varies greatly between these wavelengths

[2] These wavelengths are chosen deliberately: 850 nm is typically used in short optical links, 660 nm is the wavelength of choice for POF, and 405 nm is applied in Blu-ray disc devices.

the absorption coefficient of silicon is 0.0382 μm^{-1}. The consequence of this rather low value is that 14.8% of the impinging photon flux remains, even after traveling 50 μm through silicon (3.5). The penetration depth for this wavelength is 26.2 μm. The absorption coefficient of silicon for a wavelength of 660 nm is already 0.23 μm^{-1}, which corresponds with a penetration depth of 4.35 μm. Radiation with a wavelength of 405 nm is absorbed almost completely in the first micrometer of silicon due to its absorption coefficient of 7.55 μm^{-1} which results in a penetration depth of only 0.132 μm.

Silicon is an *indirect band gap* semiconductor. This means that the minimum energy level of the conduction band and the maximum energy level of the valence band correspond to different electron momenta. Consequently, the transition of an electron from the valence band to the conduction band requires the supply of an amount of energy equal to the band gap energy as well as the supply of a phonon[3]. Therefore, three particles are required to create an electron-hole pair, namely an electron, a photon and a phonon. This is rather an exceptional phenomenon which is the reason why the penetration depth in indirect band gap semiconductors is large, especially for photons with long wavelengths. The same is true for the radiative recombination mechanism: an electron, a photon and a phonon need to interact to enable it. This is the reason why radiative recombination is a negligible mechanism in silicon and other indirect band gap semiconductors. By contrast, a direct band gap semiconductor requires only the interaction of an electron and a photon. It is therefore much more likely in a direct semiconductor to have radiative generation and recombination. This is also the reason why the penetration depth in a direct band gap semiconductor is significantly smaller compared with that of an indirect band gap material.

To conclude, radiative recombination is a negligible mechanism in silicon and other indirect band gap semiconductors and radiative generation is the mechanism that is employed usefully to build silicon photodetectors.

Shockley-Read-Hall Recombination Ideally, the crystal structure of a semiconductor has no imperfections or impurities. The electronic states of all electrons in the structure are in that case limited to the allowed theoretical energy bands. No electron can have an energy level that has a value in between these bands. These forbidden bands are called gaps. The gap between the two highest energy bands, the valence band and the conduction band, is typically called the band gap. However, no crystal structure is perfect in reality. Imperfections and impurities are always present to some degree. These introduce electronic states with energy levels that are located within the band gap. These intermediate levels can serve as efficient generation or recombination centers. Instead of making the transition from the valence band to the conduction band and back in a single step, an electron can bridge that gap in several smaller steps. This process is called Shockley-Read-Hall generation and recombination, respectively, and is shown schematically in Fig. 3.4.

[3] A phonon is a particle characterized by the quantization of the lattice vibrations.

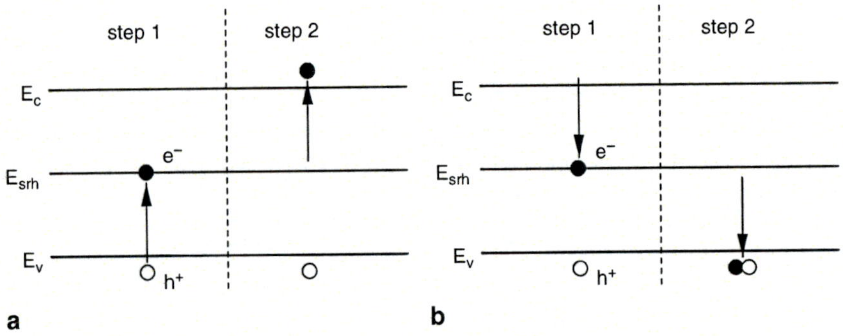

Fig. 3.4 Shockley-Read-Hall generation and recombination processes in a semiconductor with band gap E_g and a generation/recombination center at E_{srh}. **a** Shockley-Read-Hall generation. **b** Shockley-Read-Hall recombination

If a concentration of recombination centers with an energy level E_{srh} is present in the forbidden band gap, the recombination rate resulting from these centers U_{srh} can be expressed as follows for n-type and p-type semiconductors, respectively [VO00]:

$$U_{srh} = \begin{cases} (p - p_0)/\tau_{h^+} & \text{for n-type semiconductors,} \\ (n - n_0)/\tau_{e^-} & \text{for p-type semiconductors.} \end{cases} \tag{3.6}$$

where p_0 and n_0 represent the concentrations of holes and electrons in the equilibrium state while τ_{h^+} and τ_{e^-} denote the lifetimes of holes and electrons, respectively, as a result of the Shockley-Read-Hall mechanism. It should be mentioned that (3.6) **is valid only if the concentration of majority carriers is significantly higher than the** intrinsic carrier concentration n_i. In a doped semiconductor, this is however almost always the case. It can be noticed from (3.6) that the Shockley-Read-Hall recombination rate is proportional to the deviation from the equilibrium concentration of the minority carriers. The Shockley-Read-Hall recombination rate U_{srh} is positive if the minority carrier concentration is higher than the equilibrium value. In the other case, when U_{srh} is negative, carriers are generated and the process is called Shockley-Read-Hall generation. The minority carrier lifetimes τ_{h^+} and τ_{e^-} in (3.6) are inversely proportional to the concentration of recombination centers. They typically vary between 10 ms and 1 ns for concentrations of 10^{10} cm^{-3} and 10^{17} cm^{-3} recombination centers, respectively [VO00].

Until now, it has been assumed that the impurities and imperfections are located in the bulk of the semiconductor. Nevertheless, a lot more defects, impurities and imperfections are present at every semiconductor surface or interface with another material. The Shockley-Read-Hall mechanism is therefore especially of importance at these surfaces or interfaces. In contrast to the discrete electronic states associated with imperfections in the bulk, a continuum of levels between the valence band and the conduction band is present at the surface. While the bulk recombination rate (3.6) determines the number of recombined electrons and holes per unit volume and time, the surface recombination rate $U_{srh,s}$ denotes the number of recombined electrons and holes per unit of area and time [VO00]:

$$U_{srh,s} = \begin{cases} s_{h^+} \cdot (p - p_0) & \text{for n-type semiconductors,} \\ s_{e^-} \cdot (n - n_0) & \text{for p-type semiconductors,} \end{cases} \tag{3.7}$$

where s_{h^+} and s_{e^-} represent the surface recombination velocities of holes and electrons, respectively.

Both Shockley-Read-Hall recombination mechanisms represent the most important recombination processes in silicon and other indirect band gap semiconductors. Radiative recombination is, as discussed in the previous section, negligible in every indirect band gap material. Another mechanism, namely *Auger recombination*, is dominant only for highly doped semiconductors with doping concentrations above 10^{18} cm^{-3}. This is not considered further in this work because the doping concentrations that are used in photodetectors are typically below 10^{18} cm^{-3}.

3.1.2 Collecting the Generated Carriers

After being generated by impinging photons, the electrons and holes must be collected in order to produce a useful signal. Because the carriers are electrically charged, an electric field can perform this task as it forces carriers with opposite signs to drift in opposite directions. Three semiconductor devices are available that implement this concept: the photoconductor, the photodiode and the phototransistor. In all of these devices, the carriers are generated and recombined by the same mechanisms which were discussed in the previous section. They only differ in the manner of collecting the generated carriers.

A photoconductor is the most straightforward implementation of a photodetector because it comprises only a slab of semiconductor between two ohmic contacts. A voltage difference between the two contacts forces an electric current to flow through the semiconductor. Electron-hole pairs are created when light impinges on the semiconductor whereby the free electron and hole concentrations are increased above their respective equilibrium concentrations. The semiconductor conductivity is raised accordingly which is translated in turn into a larger current between the two ohmic contacts. A modulated light signal can thus be transformed into a modulated electric current. In order to maximize the current signal, the lifetime of the generated carriers needs to be longer than the time it takes to reach one of the contacts. This directly influences the conversion speed of the photoconductor because a long lifetime causes the conversion speed to be low. The conversion speed can only be increased by reducing the lifetime and consequently by decreasing the signal amplitude. This is the typical trade-off in a photoconductor.

A phototransistor is a bipolar transistor in which the base-collector junction acts as a photodiode. Although the operating principle of a photodiode is explained in the following section, it is sufficient for the moment to know that it generates a current flowing from the cathode to the anode when illuminated. Since this current can be considered as the base current, it is amplified with the current gain of the bipolar transistor. Consequently, the output signal of a phototransistor is typically large but the major drawback is the low conversion speed. This is due to the large parasitic capacitance between the base and the collector which originates from the large area that is required by the base-collector junction to capture as much light as possible.

Neither a photoconductor nor a phototransistor is used in this work as a photodetector. As already mentioned, a photoconductor is limited by a fundamental trade-off between gain and speed. Moreover, it carriers a DC current and consumes power as a result. By contrast, a phototransistor is a bipolar device which is very difficult to implement efficiently in a standard CMOS technology. Instead, a photodiode is used in this work.

Diode Fundamentals A semiconductor diode is a piece of semiconductor that is n-type doped on one side and p-type doped on the other side. The n-type side is denoted as *cathode* and it contains a lot of free electrons. The p-type side on the contrary is called the *anode* and contains a lot of free holes. The electrons and holes are thus not equally divided over the complete structure. This gives rise to electron and hole diffusion[4], which means that some electrons move from the n-type region to the p-type region and that some holes move from the p-type region to the n-type region. Due to this migration of free carriers, the n-type region is positively charged, whereas the p-type region is negatively charged. An electric field is thus established between both regions. This field provokes a drift current of electrons and holes that counterbalances the diffusion currents: electrons drift back to the n-type region and holes to the p-type region. Equilibrium is reached when the diffusion and the drift currents counterbalance each other and no net current flows from one region to the other. The energy band diagram of this situation is shown in Fig. 3.5.

As already mentioned, an electric field ξ appears between the n-type region and the p-type region originating from the charge redistribution due to diffusion. This electric field causes the electrostatic potential to vary over the structure:

$$\Psi = -\int \xi \, dx, \tag{3.8}$$

where x indicates the direction perpendicular to the junction between both regions. Because the potential varies within the diode, the energy bands of the material fluctuate accordingly. This phenomenon is denoted as band bending and it is depicted in Fig. 3.5. No band bending occurs in the neutral regions, far from the junction.

The part of the structure around the junction where the electric field manifests itself is called the *space charge region* (SCR) as it is only in this region where the structure carries a net charge: a positive charge in the n-type region and a negative charge in the p-type region. The width of the SCR, W_{SCR}, depends on the doping concentration of the n-type region N_D, the doping concentration of the p-type region

[4] Diffusion is the net movement of particles from a region of higher concentration to a region of lower concentration. Ultimately, diffusion settles to an equal distribution of these particles over the complete structure or environment.

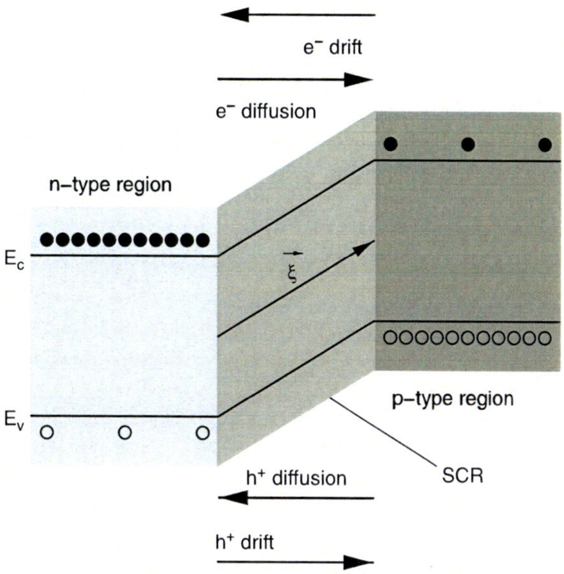

Fig. 3.5 Energy band diagram of a semiconductor diode in equilibrium; equilibrium is reached when the drift current of electrons and holes balances the diffusion current of both

N_A and the doping profile which describes the evolution of the doping concentration from the n-type region to the p-type region. If the junction is abrupt[5], W_{SCR} can be calculated as follows [VO00]:

$$W_{SCR} = \sqrt{\frac{2\varepsilon}{q} \cdot \frac{N_A + N_D}{N_A N_D} \cdot V_{bi}}, \tag{3.9}$$

where ε is the *permittivity*[6] of the semiconductor under consideration and V_{bi} is the *built-in voltage* of the diode. The latter is specified with the following formula:

$$V_{bi} = \frac{kT}{q} \cdot \ln\left(\frac{N_A N_D}{n_i^2}\right). \tag{3.10}$$

From (3.9) and (3.10) it is clear that, for a certain semiconductor, W_{SCR} depends on the doping concentrations at both sides of the diode. This dependence is plotted in Fig. 3.6 for silicon at room temperature ($T = 300$ K)[7]. Note that higher doping concentrations correspond to smaller widths for the SCR. It will be shown in the next section that W_{SCR} plays a dominant role in the performance of a photodiode. In a symmetrical diode ($N_A = N_D$), the SCR extends as far into the n-type region as

[5] An abrupt junction is a junction in which the evolution of the doping concentration from the n-type region to that of the p-type region is completed over an infinitesimally short distance around the junction.

[6] The permittivity of silicon equals $11.7 \cdot \varepsilon_0$, where ε_0 is the permittivity of vacuum ($8.85 \cdot 10^{-12}$ F/m).

[7] If no value is specified for the temperature, it is from now on assumed that $T = 300$ K.

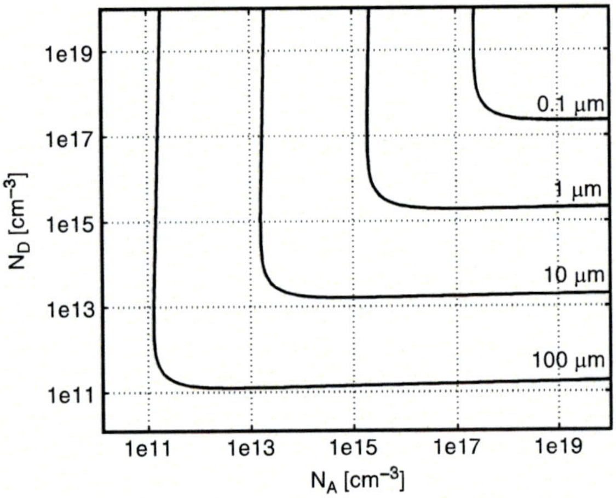

Fig. 3.6 Contour plot of W_{SCR} as a function of N_A and N_D for a silicon diode with an abrupt junction

into the p-type region. On the other hand, the SCR extends almost completely into the region with the lower doping concentration in a one-sided diode ($N_A \gg N_D$ or $N_D \gg N_A$) [VO00].

Until now, it has been assumed that no voltage is applied between the anode and cathode. Because a diode is a non-symmetrical device, its response to an applied voltage is highly non-linear. If the voltage at the anode is higher than that at the cathode ($V_{pn} > 0$), the diode is said to be forward biased, whereas it is reverse biased in the other case ($V_{pn} < 0$). The amount of bias across a diode with an abrupt junction has the following effect on the width of its SCR [VO00]:

$$W_{SCR} = \sqrt{\frac{2\varepsilon}{q} \cdot \frac{N_A + N_D}{N_A N_D} \cdot (V_{bi} - V_{pn})}. \tag{3.11}$$

The effect of a reverse bias on W_{SCR} is plotted in Fig. 3.7 for a symmetrical silicon diode. Note that the width of the SCR increases with the amount of reverse bias. By contrast, W_{SCR} decreases when the diode is forward biased.

The SCR acts as an isolating layer between two conducting layers. Increasing the bias across the diode decreases the width of the SCR by moving majority carriers into a slice of semiconductor at each side of the SCR. Decreasing the bias on the contrary increases the width of the SCR by moving majority carriers away from each side of the SCR. This behavior can be modeled as a capacitance C_j, the so-called *junction capacitance*. Its value can be calculated with the following formula in the case of an abrupt junction [Gra84]:

$$C_j = \underbrace{A \cdot \sqrt{\frac{q\varepsilon N_A N_D}{2(N_A + N_D)V_{bi}}}}_{C_{j,0}} \cdot \frac{1}{\sqrt{1 - \frac{V_{pn}}{V_{bi}}}}, \tag{3.12}$$

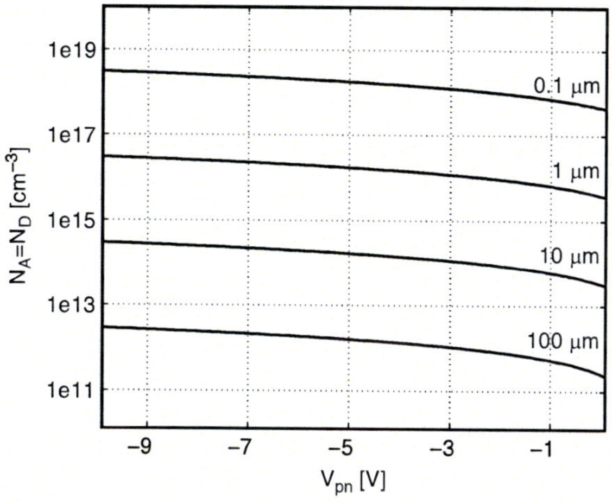

Fig. 3.7 Contour plot of W_{SCR} as a function of V_{pn} for a symmetrical silicon diode with an abrupt junction

where $C_{j,0}$ is the junction capacitance when no bias voltage is applied and A is the cross-sectional area of the junction. As can be seen in Fig. 3.8, in which a contour plot of the junction capacitance of a symmetrical silicon diode with an abrupt junction is shown, the value of the junction capacitance decreases for a decreasing doping concentration as well as for an increasing reverse bias voltage. It can be concluded from (3.12) that C_j approaches infinity when V_{pn} approaches V_{bi}. However, in that case the diode is forward biased and the current flow through the device cannot be neglected. A more detailed analysis shows that (3.12) is only valid for biasing voltages below $V_{bi}/2$ [Gra84].

As discussed before, the junction capacitance increases with an increasing forward bias voltage. Nevertheless, it can be neglected when the diode is forward biased

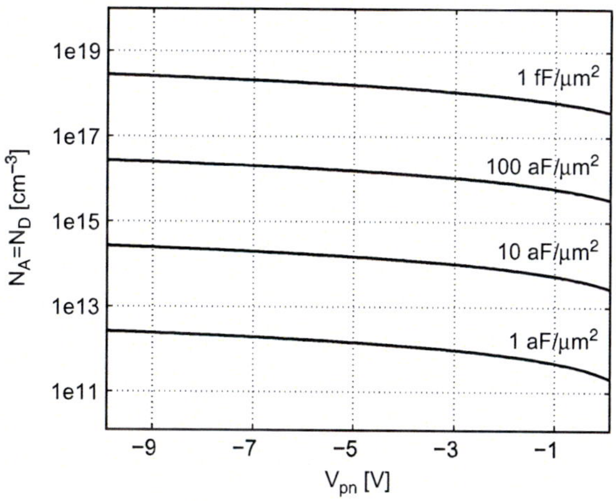

Fig. 3.8 Contour plot of C_j as a function of V_{pn} for a symmetrical silicon diode with an abrupt junction

because another capacitance, the *diffusion capacitance*, dominates in this regime. This is due to the fact that the diffusion capacitance has an exponential relationship with the bias voltage. Because a photodiode is always operated with a reverse bias voltage, this second parasitic capacitance will not be treated in more detail here.

It has been assumed until now that a semiconductor diode is composed of an n-type and a p-type region only. However, it is common practice for some applications to use diodes with three regions: an n-type region, a p-type region, and an intrinsic region sandwiched in between. Such a diode is typically called a PIN diode. The intrinsic piece of semiconductor in between the two highly doped neighboring layers acts as an extended SCR. Due to the larger SCR width, the junction capacitance of a PIN diode is significantly smaller than that of a normal diode. Furthermore, the junction capacitance of a PIN diode is almost independent of the bias voltage, in contrast to a normal diode (see Fig. 3.8). Anyway, it is not possible to integrate PIN structures in a standard CMOS technology.

Working Principle of a Photodiode If photons with enough energy impinge on a semiconductor diode, electron-hole pairs are created in the n-type region, in the p-type region, and in the SCR. The electric field within this SCR separates the electrons from the holes. In this manner, a photocurrent is generated that flows between the anode and the cathode of the diode which is then called a photodiode.

The energy band diagram of a photodiode with photons impinging on it is shown in Fig. 3.9. Electron-hole pairs that are created in the SCR are subjected to the electric field in that region. As a result, these carriers are immediately swept to the region where they are majority carriers: electrons are swept into the n-type region while holes are swept into the p-type region. This way, a drift current flowing from the n-type region to the p-type region is established. Electron-hole pairs that are generated within approximately one minority carrier diffusion length of the edge of the SCR have a high probability of diffusing towards the SCR before they recombine. The

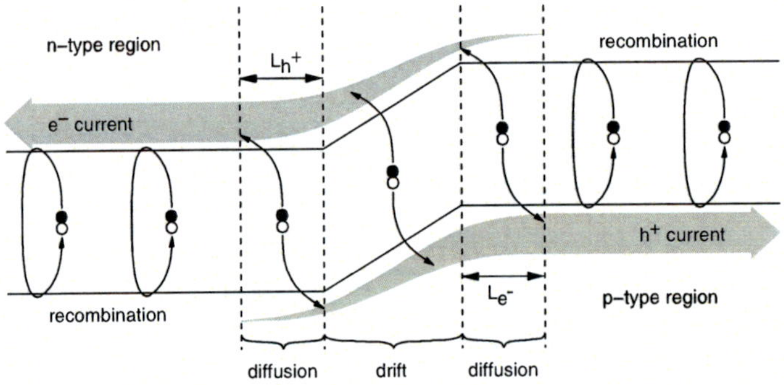

Fig. 3.9 Energy band diagram of a semiconductor photodiode on which photons are impinging

diffusion length of electrons in the p-type region (L_{e^-}) and that of holes in the n-type region (L_{h^+}) are defined by:

$$L_{e^-} = \sqrt{D_{e^-}\tau_{e^-}} = \sqrt{\mu_{e^-}\frac{kT}{q}\tau_{e^-}}, \tag{3.13}$$

$$L_{h^+} = \sqrt{D_{h^+}\tau_{h^+}} = \sqrt{\mu_{h^+}\frac{kT}{q}\tau_{h^+}}, \tag{3.14}$$

respectively, where D_{e^-} and D_{h^+} are the diffusion constants of electrons and holes, respectively, μ_{e^-} and μ_{h^+} represent the electron and hole mobilities and τ_{e^-} and τ_{h^+} denote the average lifetimes of an electron and a hole. The diffusion length can be considered as the average distance that can be travelled by a minority carrier before being annihilated via a recombination process. Electrons and holes in these neutral regions are not subjected to an electric field. Consequently, their movement is governed by the diffusion mechanism. However, from the moment these carriers reach the edge of the SCR they are immediately swept away by the present electric field to the other side of the SCR. In this way, they generate a diffusion current that flows from the n-type region to the p-type region. Electron-hole pairs that are created far from the SCR recombine before they can diffuse towards the SCR. Consequently, these carriers are never collected and they do not contribute to the photocurrent.

The current through a semiconductor diode has an exponential relationship with the voltage across it, namely:

$$I_{pn} = I_s \cdot \left(e^{\frac{qV_{pn}}{kT}} - 1\right), \tag{3.15}$$

where I_s is the saturation current of the diode, which typically has an extremely small value. It depends on the area of the junction, the doping concentrations, the doping profiles, and the carrier mobilities. If V_{pn} is positive, which means that the p-type region is biased with a higher voltage than the n-type region, the diode is forward biased and the current through the device increases exponentially with an increasing bias voltage. The diode is reverse biased if V_{pn} is negative. The current through the device saturates at I_s and flows from the n-type region to the p-type region. This behavior is shown in Fig. 3.10.

If the diode is illuminated with a certain optical power, the current–voltage characteristic is shifted downwards, as demonstrated in Fig. 3.10. The size of the shift is determined by the number of electron-hole pairs that are created and consequently by the number of photons that impinge on the diode. This working principle is identical to the one that is used in a solar cell which converts solar energy into electrical energy. However, the objective of a solar cell is to deliver power whereas a photodiode always consumes power. The difference is only made in the manner they are operated in a circuit. As depicted in Fig. 3.10, a photodiode is biased with a reverse voltage so that its operating point is always in the third quadrant of the coordinate system. Considering the current that flows from the cathode to the anode and the voltage that is higher at the cathode, the photodiode consumes an amount of power equal to

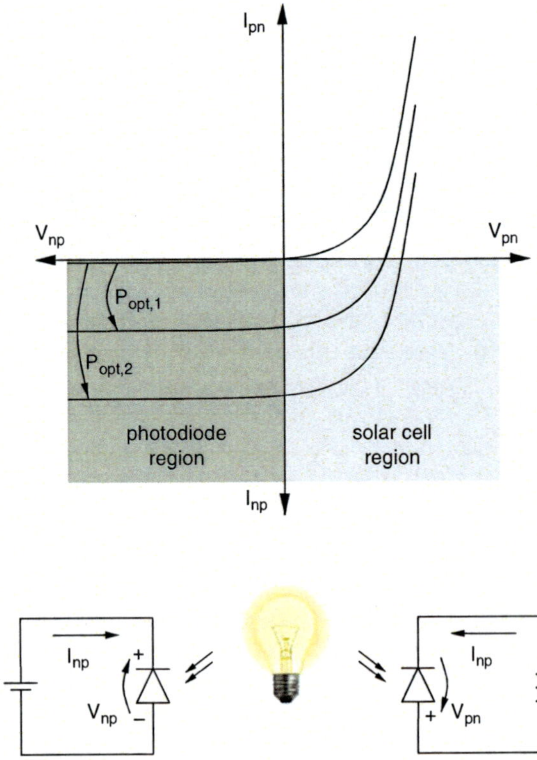

Fig. 3.10 Current–voltage characteristic of a semiconductor diode in three different situations: non-illuminated, illuminated with a power $P_{opt,1}$ and illuminated with a power $P_{opt,2} > P_{opt,1}$

$I_{np}V_{np} = I_{pn}V_{pn}$. On the contrary, a solar cell is not biased by an external voltage and is as a result always operated in the fourth quadrant of its current-voltage characteristic. The solar cell consumes an amount of power equal to $I_{np}V_{pn} = -I_{pn}V_{pn}$ which is negative and therefore means that it delivers power to the connected load resistor.

The electrical models of an illuminated and a non-illuminated photodiode are shown in Fig. 3.11. The parasitic capacitance of a photodiode C_{pd} is essentially the junction capacitance because the diffusion capacitance of a reverse biased junction is significantly smaller. The series resistance R_{pd} models the path and contact resistances. As it is in series with C_{pd} and introduces a time constant $R_{pd}C_{pd}$ as a result, it needs to be minimized. This can be accomplished by adding many contacts to the anode and the cathode; this in turn compromises the efficiency of the photodiode since metal contacts reflect the incident light. The current through an illuminated semiconductor junction flows from the n-type region to the p-type region and varies linearly with the impinging light power (Fig. 3.11a). The reverse current through the junction is reduced to the saturation current I_s when the photodiode is non-illuminated (Fig. 3.11b). If the light signal is periodically switched on and off, which is the case in an optical communication system transmitting digital data, the electrical model of the photodiode switches between those shown in Fig. 3.11a and b.

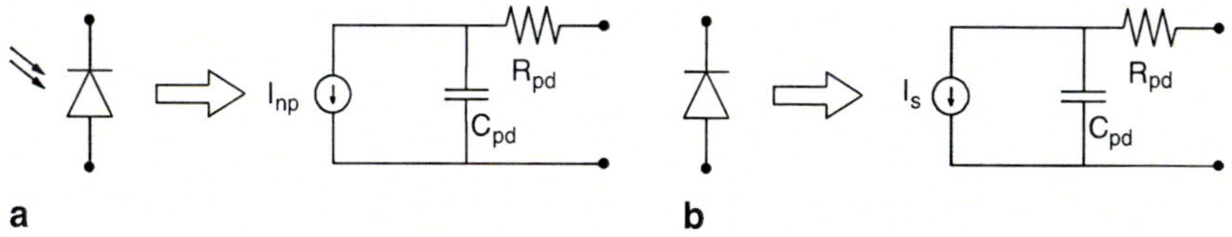

Fig. 3.11 Electrical model of an illuminated and a non-illuminated photodiode with parasitic capacitance C_{pd} and series resistance R_{pd}. **a** An illuminated photodiode conducts a reverse current I_{np}. **b** A non-illuminated photodiode conducts the saturation current I_s

3.2 Photodiodes in CMOS

Photodiodes can be integrated in silicon CMOS processes. The choice of this kind of technology makes the design of a photodiode very challenging. The first reason for this is the fact that silicon is used as the base material. As discussed in the previous section, silicon is an indirect band gap semiconductor. Consequently, its absorption coefficient is relatively low, which means that photons are only absorbed gradually by the material. As the penetration depth of light with a wavelength of 850 nm in silicon is approximately 26.2 μm, the generated electron-hole pairs are distributed over a large depth in the material. Consequently, it is difficult to collect all these carriers to generate a photocurrent; this typically leads to a low responsivity due to the recombination of many of the generated carriers.

The second reason for the difficulties that arise when integrating the photodiode in CMOS is the limited freedom there is in realizing a junction in this technology. In the schematic overview of a typical CMOS process in Fig. 1.6, it can be seen that basically five different layers can be manufactured in it: a p-type substrate, a p-well, an n-well, a p+ region and an n+ region. As the substrate is always p-type, only three different junctions can be realized in CMOS; these are shown in Fig. 3.12. The depths and doping concentrations of these layers are all technology dependent and can therefore not be altered. The design of a photodiode can hence be simplified by choosing between the three feasible junctions and selecting its dimensions.

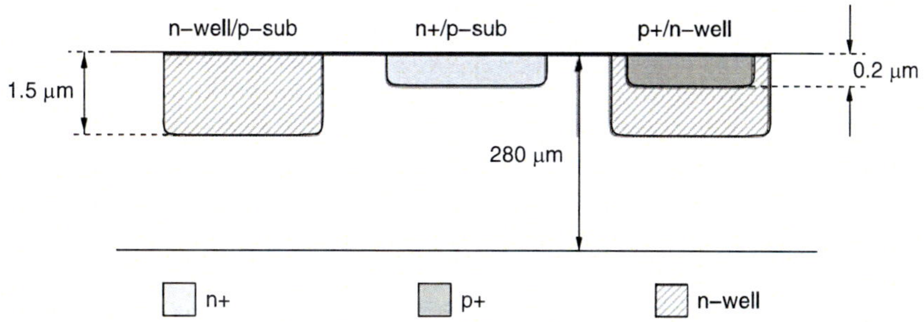

Fig. 3.12 Feasible photodiode junctions in a standard CMOS process with some typical dimensions

Table 3.1 Calculated W_{SCR} for the three feasible CMOS junctions for two different bias voltages

	W_{SCR}	
	$V_{pn} = -1$ V	$V_{pn} = 0$ V
n-well/p-sub	1.73 μm	1.11 μm
n+/p-sub	1.78 μm	1.19 μm
p+/n-well	0.160 μm	0.112 μm

The exact doping concentration of the available layers in a specific technology is typically not disclosed by the technology provider due to confidentiality issues. However, to do some basic calculations, the following values are assumed for now:

$$N_{p\text{-}sub} = 7 \cdot 10^{14}\ \text{cm}^{-3},$$

$$N_{n\text{-}well} = N_{p\text{-}well} = 10^{17}\ \text{cm}^{-3},$$

$$N_{n+} = N_{p+} = 10^{19}\ \text{cm}^{-3}.$$

It is also assumed that the junctions are abrupt. The SCR width of every available junction can then be calculated using (3.11) of which the results are summarized in Table 3.1. Two important conclusions can be drawn from these results:

1. The width of the SCR is determined mainly by the region with the lower doping concentration;
2. The width of the SCR is very small compared with the penetration depth of visible light.

The n-well/p-sub and the n+/p-sub photodiodes are simple single junctions with an approximately equal W_{SCR}. By contrast, the p+/n-well photodiode has an intrinsic second junction between the n-well and the substrate. Instead of using only the p+/n-well, both junctions can be used in parallel by connecting the substrate electrically to the p+ region. In that case, the effective W_{SCR} is the sum of the separate widths: $0.112 + 1.11\ \mu\text{m} = 1.22\ \mu\text{m}$. This is only slightly greater than the width of a single n-well/p-sub or n+/p-sub junction. If only the p+/n-well junction is used, all carriers that are generated in the substrate are lost as they never appear in the photodiode response. As discussed in Sect. 3.2.2, this can be employed to increase the speed of the photodiode.

3.2.1 *Magnitude and Speed of the Photocurrent*

As shown in the previous chapter, the responsivity of the photodiode determines the sensitivity of an optical receiver. The speed or bandwidth of the photodiode limits the highest attainable bit rate. Both are important figures during the design of a high-speed optical receiver and are therefore discussed in more detail here. It is assumed from now on that an n-well/p-sub photodiode is used. It is also assumed that the photodiode is a one-dimensional structure of which the important dimensions are defined in Fig. 3.13.

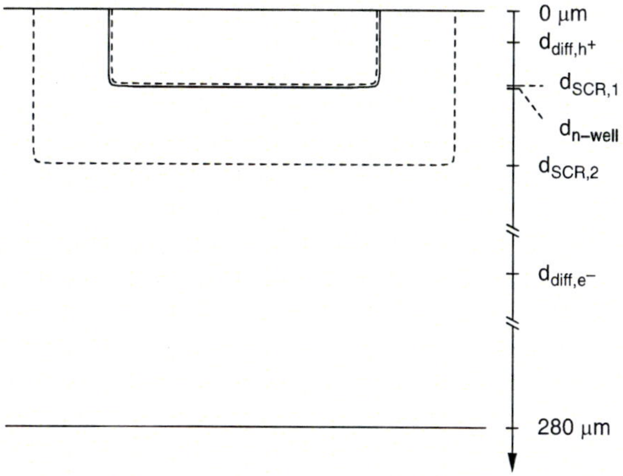

Fig. 3.13 Important dimensions of an n-well/p-sub photodiode

Magnitude The responsivity of a photodiode is defined as the magnitude of the photocurrent, which is normalized to the optical input power. Consequently, knowledge of this magnitude for an optical input power of 1 W reveals the value of the responsivity. The total photocurrent is composed of three components, namely the drift component, the diffusion component of holes in the n-type region, and the diffusion component of electrons in the p-type region:

$$I_{pd} = I_{drift} + I_{diff,h^+} + I_{diff,e^-}. \tag{3.16}$$

To determine the magnitude of the drift current, the number of photons that is absorbed in the SCR per unit time and area, ϕ_{drift}, should be known. As it is the difference in photon flux between both SCR boundaries, ϕ_{drift} can be calculated with (3.5) as follows:

$$\begin{aligned} \phi_{drift} &= \Phi(d_{SCR,1}) - \Phi(d_{SCR,2}), \\ &= \Phi_0 \cdot e^{-\alpha d_{SCR,1}} - \Phi_0 \cdot e^{-\alpha d_{SCR,2}}, \\ &= \Phi_0 \cdot \left(e^{-\alpha d_{SCR,1}} - e^{-\alpha d_{SCR,2}}\right), \end{aligned} \tag{3.17}$$

where Φ_0 represents the photon flux at the surface of the substrate. As discussed in Sect. 3.1.2, the SCR extends almost completely into the region with the lower doping concentration. Therefore, it is reasonable to assume that the upper border of the SCR in the n-well is at the same depth in the substrate as the junction border itself ($d_{SCR,1} = d_{n\text{-}well}$). In a typical CMOS process, this junction border is at a depth of approximately 1.5 μm (Fig. 3.12). The other border of the SCR is localized in the p-type substrate and is at a depth $d_{SCR,2}$, which is defined as follows:

$$d_{SCR,2} = d_{SCR,1} + W_{SCR}. \tag{3.18}$$

With W_{SCR} for an n-well/p-sub junction being given in Table 3.1, $d_{SCR,2}$ can be calculated to be 2.61 μm or 3.23 μm for bias voltages of 0 V and −1 V, respectively.

For light with a wavelength of 850 nm and a corresponding absorption coefficient in silicon of 0.0382 μm^{-1}, (3.17) reveals that ϕ_{drift} varies between 3.92 and 6.04% of Φ_0 depending on the biasing condition.

The hole diffusion current originates from carriers that are generated in the n-well within approximately one diffusion length of the upper border of the SCR. The number of photons per unit time and area that is absorbed in this region can be expressed by the following formula:

$$\phi_{diff,h^+} = \Phi_0 \cdot (e^{-\alpha d_{diff,h^+}} - e^{-\alpha d_{SCR,1}}), \tag{3.19}$$

where d_{diff,h^+} is the depth beyond which all generated carriers are assumed to reach the SCR before recombining. Its value can be calculated as follows:

$$d_{diff,h^+} = d_{SCR,1} - L_{h^+}, \tag{3.20}$$

where L_{h^+} is the diffusion length of holes in the n-well. To obtain the value of L_{h^+}, (3.14) reveals that both the hole mobility μ_{h^+} and the hole lifetime τ_{h^+} in the n-well should be known. For a doping concentration of 10^{17} cm^{-3}, which is assumed to be the doping concentration of the n-well, μ_{h^+} is approximately 300 cm^2/Vs [Zim04]. The lifetime of a hole in the n-well depends on the number of recombination centers and therefore on the purity of the crystal. Although the value of τ_{h^+} is not known precisely as it is a technology parameter which is not disclosed by the manufacturer, it is often assumed to be approximately 1 μs [Her07, Zim04] for a silicon CMOS process. For that reason, this value is used in this work. For such a long lifetime, L_{h^+} can be calculated to be 27.8 μm. This forces d_{diff,h^+} to become negative (3.20), which means that all carriers that are generated in the n-well are also collected by the SCR after a certain amount of time. Formula (3.19) can now be simplified to the following:

$$\phi_{diff,h^+} = \Phi_0 \cdot (1 - e^{-\alpha d_{SCR,1}}). \tag{3.21}$$

If $d_{SCR,1}$ is again supposed to be 1.5 μm, ϕ_{diff,h^+} is 5.57% of Φ_0 for light with a wavelength of 850 nm, regardless of the bias voltage.

Clearly, most of the photons are absorbed in the p-type substrate where the electron diffusion current originates. This electron diffusion current is formed by all the carriers that are generated between the lower border of the SCR and d_{diff,e^-}. The number of photons per unit time and area that is absorbed in this region can be expressed as follows:

$$\phi_{diff,e^-} = \Phi_0 \cdot (e^{-\alpha d_{SCR,2}} - e^{-\alpha d_{diff,e^-}}), \tag{3.22}$$

where d_{diff,e^-} is the depth for which all carriers that are generated above it are assumed to reach the SCR before recombining. Its value is determined by the diffusion length of electrons in the substrate:

$$d_{diff,e^-} = d_{SCR,2} + L_{e^-}, \tag{3.23}$$

where L_{e^-} is the diffusion length of electrons in the p-type substrate. The electron mobility μ_{e^-} in the substrate with a doping concentration of $7 \cdot 10^{14}$ cm^{-3} is 1300 cm^2/Vs [Zim04]. Similar to the hole lifetime, the electron lifetime is assumed to be 1 μs. With (3.13), L_{e^-} can in that case be determined to be 58.0 μm. For bias voltages across the junction of 0 V and −1 V, (3.23) shows that d_{diff,e^-} varies between 60.6 and 61.2 μm, respectively, because $d_{SCR,2}$ changes between 2.61 and 3.23 μm as derived before. Therefore, the total number of absorbed photons per unit time and area ϕ_{diff,e^-} varies from 80.6 to 78.7% of Φ_0 for bias voltages of 0 V and −1 V, respectively, as predicted by (3.22).

All of the photons that are absorbed in the substrate at a depth below d_{diff,e^-} simply recombine and never contribute to the photodiode current. The total number of lost photons per unit time and area ϕ_{lost} can be calculated as follows:

$$\phi_{lost} = \Phi_0 \cdot e^{-\alpha d_{diff,e^-}}. \tag{3.24}$$

For bias voltages of 0 V and −1 V, ϕ_{lost} varies between 9.88 and 9.65% of Φ_0, respectively.

In Table 3.2, the results from above are summarized and compared with the situation where light with a wavelength of 660 nm is used instead of the more typical 850 nm. It can be noticed immediately that more photons are absorbed in the SCR and the n-well due to the higher absorption coefficient in silicon of light with a shorter wavelength. This has a severe impact on the speed of the photodiode, as demonstrated in the next section.

Under the assumption that all of the absorbed photons are converted into electron-hole pairs, the relative distribution of the total photocurrent between the three constituent components in (3.16) is given in Table 3.3. For a fixed number of impinging photons, the magnitude of I_{pd} is larger if light with a wavelength of 660 nm is

Table 3.2 Relative distribution of the absorbed photons in an n-well/p-sub photodiode between the different regions for a wavelength of 850 nm and 660 nm

	λ = 850 nm		λ = 660 nm	
	$V_{pn} = -1$ V	$V_{pn} = 0$ V	$V_{pn} = -1$ V	$V_{pn} = 0$ V
ϕ_{drift}/Φ_0	6.04%	3.92%	23.2%	16.0%
$\phi_{diff,h^+}/\Phi_0$	5.57%	5.57%	29.2%	29.2%
$\phi_{diff,e^-}/\Phi_0$	78.7%	80.6%	47.6%	54.9%
ϕ_{lost}/Φ_0	9.65%	9.88%	0.00%	0.00%

Table 3.3 Relative distribution of the total photocurrent of an n-well/p-sub photodiode between the different current components for wavelengths of 850 nm and 660 nm

	λ = 850 nm		λ = 660 nm	
	$V_{pn} = -1$ V	$V_{pn} = 0$ V	$V_{pn} = -1$ V	$V_{pn} = 0$ V
I_{drift}/I_{pd}	6.69%	4.35%	23.2%	16.0%
$I_{diff,h^+}/I_{pd}$	6.17%	6.18%	29.2%	29.2%
$I_{diff,e^-}/I_{pd}$	87.1%	89.5%	47.6%	54.9%

used because no photons are lost deep in the substrate. However, for a fixed optical power this is no longer true as photons with a short wavelength have more energy than photons with a long wavelength. This can be expressed mathematically as follows:

$$I_{pd} = q \cdot \frac{\phi_{drift} + \phi_{diff,h^+} + \phi_{diff,e^-}}{\Phi_0} \cdot \frac{P_{opt}}{h\nu} \tag{3.25}$$

If light with a power of 1 W and a wavelength of 850 nm is used to generate I_{pd}, its value can be calculated to be 0.618 A or 0.619 A for bias voltages of 0 V and −1 V, respectively. The theoretical responsivity of a silicon n-well/p-sub photodiode for light with a wavelength of 850 nm is thus 0.618 A/W or 0.619 A/W, depending on the bias voltage. For 1 W of light with a wavelength of 660 nm, I_{pd} has a value of 0.532 A, regardless of the bias voltage. The responsivity of a silicon n-well/p-sub photodiode for light with a wavelength of 660 nm is thus 0.532 A/W, which is indeed lower than the responsivity for light with a longer wavelength.

The theoretical derived responsivities differ from the actual responsivities for several reasons. Firstly, the effect of surface recombination has been neglected because the relevant technological parameters are not disclosed by the manufacturer. Since many recombination centers are present at the substrate surface, a significant number of generated electron-hole pairs recombine in the region close to the surface, especially for light with a short wavelength. Secondly, it has been assumed that all photons are converted into electron-hole pairs, while in reality some of them are converted in phonons and, as a consequence, simply heat the material. Thirdly, the exact doping profiles and carrier lifetimes are unknown because, again, this information is not disclosed by the manufacturer. Finally, the stack of dielectric layers on top of the substrate reflects some of the impinging light. The latter issue is treated in detail in Sect. 3.2.4.

Speed In the previous section, it was shown that the three current contributions, I_{drift}, I_{diff,h^+} and I_{diff,e^-} of the total photocurrent I_{pd} have significantly different magnitudes. In addition, they show a significant difference in speed. This issue is discussed in this section.

Electrons and holes that are generated in the SCR are subjected to the electric field ξ that is present in the region. Under the assumption that ξ is constant over the SCR, its magnitude can be quantified as follows:

$$|\xi| = \frac{V_{bi} - V_{pn}}{W_{SCR}}. \tag{3.26}$$

An electron and a hole that drift due to this electric field obtain a velocities of v_{e^-} and v_{h^+}, respectively, that are given by:

$$v_{e^-} = -\mu_{e^-} \cdot |\xi|, \tag{3.27}$$

$$v_{h^+} = \mu_{h^+} \cdot |\xi|. \tag{3.28}$$

The mobilities μ_{e^-} and μ_{h^+} depend on the doping concentration and also on the magnitude of the electric field. For a high reverse bias voltage, and thus a large electric field, the carrier velocities saturate at $v_{e^-,sat}$ and $v_{h^+,sat}$. Both equal approximately 10^7 cm/s. In reality, however, the electric field is not constant over the complete SCR [VO00]. Rather, it needs to be considered as a function of d, the depth in the substrate. As a result, the velocities of the electrons and holes also depend on d. The transit time of an electron or hole is the time it takes them to reach one of the edges of the SCR and it is therefore a function of the carrier speed, v_{e^-} or v_{h^+}, and W_{SCR}. The average transit times τ_{t,e^-} and τ_{t,h^+} of an electron and hole through the SCR define the speed of the electron and hole drift currents. The speed, expressed in terms of the -3 dB frequency, of the electron and hole drift currents can be calculated respectively with the following formulas [Rad06]:

$$f_{drift,e^-} = \frac{2.4}{2\pi \tau_{t,e^-}}, \tag{3.29}$$

$$f_{drift,h^+} = \frac{2.4}{2\pi \tau_{t,h^+}}. \tag{3.30}$$

The values of τ_{t,e^-} and τ_{t,h^+} are 8.24 ps and 36.3 ps, respectively, in a 180 nm CMOS technology with doping concentrations that are comparable to the ones used in this chapter [Rad06]. Hence, these values are also applied in this work, resulting in electron and hole drift current bandwidths of 46.4 GHz (3.29) and 10.5 GHz (3.30), respectively.

In contrast to the drift current, the hole diffusion current which originates in the n-well depends strongly on the photodiode layout. In [Gen01], a fingered n-well layout is considered to verify the speed behavior of the hole diffusion current. Such a photodiode layout is shown in Fig. 3.14. It is concluded that the -3 dB frequency of this current component is determined by the wavelength λ of the applied light, the hole diffusion constant D_{h^+}, the n-well depth $d_{n\text{-}well}$, the n-well finger width $w_{n\text{-}well}$ and the hole diffusion length in the n-well L_{h^+} [Gen01, Rad06]:

$$f_{diff,h^+} \approx \sqrt[3]{\frac{\lambda}{850\text{ nm}}} \cdot \frac{\pi D_{h^+}}{2} \cdot \left(\left(\frac{1}{2d_{n\text{-}well}}\right)^2 + \left(\frac{1}{w_{n\text{-}well}}\right)^2 + \left(\frac{1}{L_{h^+}}\right)^2 \right). \tag{3.31}$$

As calculated before, the hole diffusion length L_{h^+} is 27.8 µm in the n-well. This is significantly larger than $d_{n\text{-}well}$, which is typically 1.5 µm in a CMOS process. Therefore, its effect on f_{diff,h^+} can be neglected. The n-well finger width $w_{n\text{-}well}$ should be minimized in order to maximize f_{diff,h^+} (3.31). However, the minimum width of an n-well is limited to approximately 1 µm in a contemporary CMOS process. With a

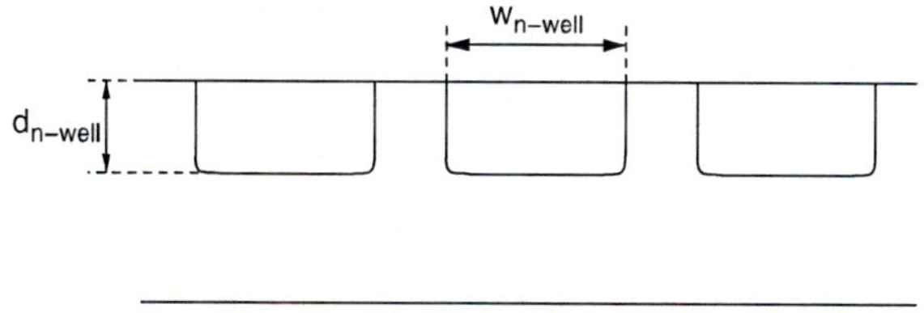

Fig. 3.14 Fingered n-well layout of an n-well/p-sub photodiode

hole mobility of 300 cm^2/Vs, the hole diffusion constant D_{h^+} results in 776 mm^2/s (3.14). If the minimum n-well width of 1 μm is used in the photodiode layout, (3.31) results in a hole diffusion current bandwidth of 1.36 GHz for light with a wavelength of 850 nm. If the n-well width is increased to 10 μm, f_{diff,h^+} drops to 149 MHz for light with the same wavelength, which is significantly lower. Holes have to diffuse further on average before reaching the border of the SCR for this larger n-well width; this explains the lower bandwidth of such a layout. In reality, the choice of $w_{n\text{-}well}$ is between speed and responsivity because the responsivity decreases with an increasing number of fingers due to layout overhead.

From (3.31), it follows that the bandwidth of the hole diffusion current decreases if light with a shorter wavelength is applied. In that case, more electron-hole pairs are generated close to the surface due to the larger absorption coefficient. This is detrimental for the speed because a hole needs to travel further on average to reach the SCR. If light with a wavelength of 660 nm is used, the bandwidth of the hole diffusion current decreases to 1.05 GHz and 116 MHz for a 1 μm and a 10 μm wide n-well, respectively.

The last but most dominant contribution to the total photodiode current is the electron diffusion current that originates in the substrate. The bandwidth of this current can be estimated with the following formula [Gen01]:

$$f_{diff,e^-} = \left(4 - \sqrt{7}\right) \cdot \frac{(\alpha L_{e^-} + 1)^2}{2\pi \tau_{e^-}}. \tag{3.32}$$

In contrast to the bandwidth of the hole diffusion current (3.31), the bandwidth of the electron diffusion current is only defined by material and technology specific constants. This means that f_{diff,e^-} is constant and cannot be altered at design time. If light with a wavelength of 850 nm is used, the absorption coefficient α equals 0.0382 μm^{-1}. The electron lifetime is again assumed to be 1 μs and the electron diffusion length in the p-type substrate is 58.0 μm, as before. In these circumstances, f_{diff,e^-} can be calculated to be equal to 2.23 MHz. By contrast, if light with a wavelength of 660 nm is applied, the bandwidth increases to 44.3 MHz due to a significantly larger absorption coefficient of 0.230 μm^{-1}.

Although the magnitudes of the three current components change with the bias voltage (Table 3.3), the bandwidths of the two diffusion currents do not. On the contrary, the bandwidth of the drift current rises with an increasing bias voltage up to the point where the carriers reach their saturation velocities. However, because the speed of the drift component is higher than 10 GHz, which is always significantly higher than the bandwidth of the other components, f_{drift} is assumed to be infinite for the remainder of the derivation. The approximate bandwidths of the different current components of a CMOS photodiode when illuminated with 850 nm light are summarized in Table 3.4.

The magnitude of the frequency response of an n-well/p-sub photodiode is shown in Fig. 3.15 for light with a wavelength of 850 nm and a bias voltage of −1 V. The electron diffusion current dominates the photodiode response at low frequencies due to the large amount of photons absorbed in the substrate. Therefore, the −3 dB

Table 3.4 Approximate bandwidth of the three components of the current of an n-well/p-sub photodiode when illuminated with 850 nm light; the bandwidth of the hole diffusion currents depends on the photodiode layout

f_{drift}	> 10GHz
f_{diff,h^+}	150 MHz–1.5 GHz
f_{diff,e^-}	2.5 MHz

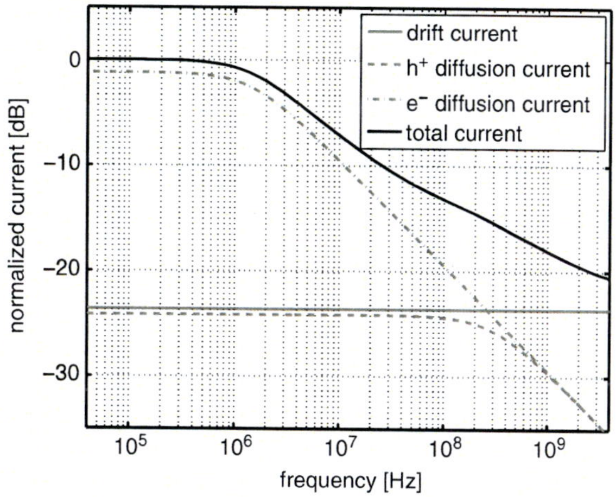

Fig. 3.15 Calculated magnitude of the photocurrent and its constituent components as a function of frequency for light with a wavelength of 850 nm and a photodiode bias voltage of −1 V

frequency of the total photocurrent is equal to that of the electron diffusion current, namely about 2.5 MHz, which is also concluded in [Gen01, Rad06]. The drift current takes over at very high frequencies thanks to its fast behavior. As a result, roll-off can be noticed for intermediate frequencies. The response of both diffusion currents shows a roll-off of 10 dB/decade [Gen01, Rad06]. The resulting roll-off of the total photodiode current is then between 4 and 10 dB/decade due to the summation of the three currents [Rad06].

3.2.2 *Speed Improvement Techniques*

In order to be able to design a high-speed optical receiver with an integrated photodiode in CMOS, the speed of the photodiode needs to be increased significantly above the calculated value of 2.5 MHz. Four techniques to realize this are discussed in this section. All of these techniques try to eliminate the detrimental effect of the electron diffusion current on the speed of a CMOS photodiode.

Using a Shorter Wavelength The most straightforward technique to increase the speed of an n-well/p-sub photodiode is to make use of light with a shorter wavelength. Fewer photons are then generated deep in the substrate and the bandwidth of the electron diffusion current rises as predicted by (3.32). The drawback of this technique is the intrinsically lower responsivity when using a shorter wavelength which is due to the fact that a photon can only generate a single electron-hole pair (3.25). In reality, the wavelength is determined by the application domain: 850 nm light in short-haul communication systems, 660 nm light in systems that use POF, 405 nm in the pickup unit of a Blu-ray disc, etc.

P+/n-well Photodiode with Guard As previously discussed, a p+/n-well diode has a second junction between the n-well and the p-type substrate. This second junction can be applied to shield the slow electron diffusion current from the photodiode response [Woo98, Her07]. The total photocurrent[8] again has three components: a drift current of carriers that are generated in the SCR of the p+/n-well junction, an electron diffusion current from the p+ region and a hole diffusion current from the n-well.

The magnitude of the drift current of a p+/n-well photodiode is smaller than that of an n-well/p-sub photodiode due to the shallower SCR (Table 3.1). The speed of this component is higher for the same reason. The electron diffusion current from the p+ region is small due to the low depth of the junction which is typically about 0.2 μm (Fig. 3.12). Consequently, only a few photons are absorbed in this region, as determined by Lambert-Beer's Law (3.5). Because an electron generated in the p+ region needs to diffuse over a distance of at most 0.2 μm to reach the edge of the SCR, the speed of this diffusion component is also high. Holes that are generated in the n-well diffuse with a more or less equal probability towards the SCR of the p+/n-well junction and that of the n-well/p-sub junction. While the former carriers contribute to the photocurrent, the latter do not. The hole diffusion current is therefore smaller than the hole diffusion current of an n-well/p-sub photodiode. For the same reason, its speed is higher compared to that of an n-well/p-sub photodiode because the average distance for a hole to reach a SCR is lower.

From the discussion above it should be clear that only the carriers that are generated in the upper 0.85 μm of the photodiode contribute to the photocurrent. If light with a wavelength of 850 nm is used, this means that 96.8% of all impinging photons are lost. This loss decreases to 82.2% if 660 nm light is applied and to only 0.163% for 405 nm light. Therefore, it can be concluded that this type of photodiode is an ideal choice provided that the wavelength of the light used is short enough.

The Differential Photodiode The third technique to increase the speed of a CMOS photodiode is to modulate the impinging light spatially in order to differentiate the fast components from the slow components [Gen01, Her07]. This can be done by realizing a photodiode with a layout as depicted in Fig. 3.16. Such a *differential photodiode* is actually composed of an alternating pattern of illuminated junctions and non-illuminated junctions, covered by metal. All illuminated junctions are connected to each other in order to form the illuminated photodiode whereas all non-illuminated junctions together form the non-illuminated photodiode.

---

[8] It is assumed in this work that the photocurrent is defined as the current flowing through the contact of the p+ region. In that case, the n-well contact is biased at a fixed voltage, typically the supply voltage.

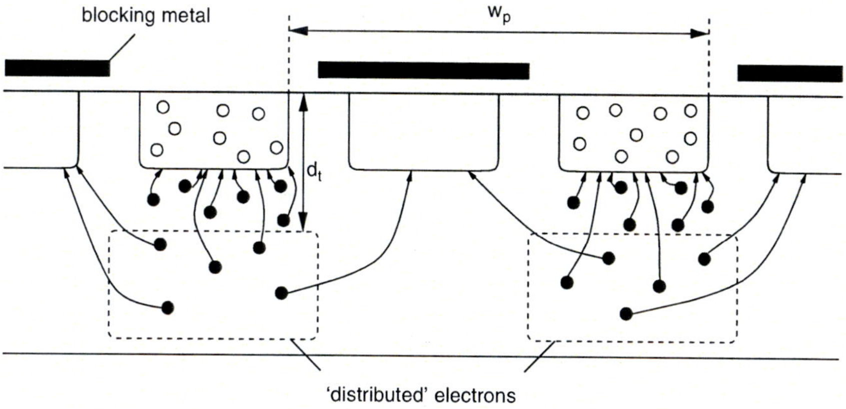

Fig. 3.16 Layout and working principle of a differential n-well/p-sub photodiode

When light impinges on this differential photodiode, electron-hole pairs are only created in the illuminated photodiode. Again, three current components are present in its response: a drift component, a hole diffusion component, and an electron diffusion component. Electrons that are generated deep below the SCR of the illuminated photodiode have approximately equal probabilities to reach the junction of an illuminated junction or that of a non-illuminated junction [Gen01]. The electron diffusion current is thus split into three parts: a relatively fast component which only appears in the response of the illuminated photodiode and two slow components originating from carriers that are generated deep in the substrate. One of these slow components appears in the response of the illuminated photodiode while the other one shows up in the response of the non-illuminated photodiode. Because the low speed of a normal n-well/p-sub photodiode is due to these deeply generated carriers, a fast response, $I_{pd,d}$, is obtained by subtracting the current of the non-illuminated photodiode, $I_{pd,ni}$, from the current of the illuminated photodiode, $I_{pd,i}$, as both have a slow component with the same magnitude. This can be expressed as follows:

$$I_{pd,d} = \underbrace{I_{drift} + I_{diff,h^+} + I_{diff,e^-,fast} + \frac{I_{diff,e^-,slow}}{2}}_{I_{pd,i}} - \underbrace{\frac{I_{diff,e^-,slow}}{2}}_{I_{pd,ni}} \tag{3.33}$$

The depth d_t at which the generated electrons start to be equally distributed between the illuminated junctions and the non-illuminated junctions depends on the periodicity of the photodiode layout, w_p: the smaller w_p, the smaller this transition depth d_t. In that case, more electrons are distributed, which results in a larger $I_{diff,e^-,slow}$ and therefore a smaller $I_{diff,e^-,fast}$. Consequently, the magnitude of the differential photocurrent decreases if w_p is decreased (3.33), which is also demonstrated in [Her07]. By contrast, the speed of $I_{diff,e^-,fast}$ increases if w_p is decreased as the maximum diffusion distance of an electron in the substrate is decreased accordingly. The following formula can be applied to estimate the speed of $I_{diff,e^-,fast}$ [Gen01]:

$$f_{diff,e^-} = 2\sqrt{3}\pi D_{e^-} \cdot \left(\left(\frac{1}{w_p} \right)^2 + \left(\frac{1}{L_{e^-}} \right)^2 \right). \tag{3.34}$$

With an electron mobility of 1300 cm^2/Vs, the electron diffusion constant D_{e^-} is 3360 mm^2/s (3.13). For periodicities of 10 μm and 5 μm, (3.34) results in a bandwidths of 377 MHz and 1.47 GHz, respectively, which is indeed a huge improvement compared with the bandwidth of 2.23 MHz for the electron diffusion current of a non-differential n-well/p-sub photodiode. Consequently, the bandwidth of $I_{pd,d}$ is increased by at least the same amount provided that the electron diffusion component remains the dominant component in the total photocurrent. Similar conclusions are drawn in [Her07] where it is derived that the bandwidth of a differential photodiode in a 180 nm CMOS technology is between 1 and 4.4 GHz for a periodicity that varies between 8.9 μm and 4.7 μm. The simulated responsivity in [Her07] drops, just as reasoned above, from 0.052 to 0.034 A/W if the periodicity is decreased between the same values as more electrons are distributed between the illuminated and the non-illuminated photodiodes.

Although the differential photodiode is able to increase the speed of an integrated photodiode from the MHz-range to the GHz-range, this technique has one major drawback, namely a low responsivity. This is mainly due to two reasons:

1. The deeper generated electron-hole pairs are removed from the output signal;
2. Half of the photodiode surface is covered with metal and therefore half of the available photons are reflected.

The Speed-Enhanced Photodiode The fourth technique, which was developed during the course of the research that led to this book, combines the responsivity advantage of a normal n-well/p-sub photodiode with the speed advantage of a differential n-well/p-sub photodiode. The layout and working principle of this *speed-enhanced photodiode* are shown in Fig. 3.17.

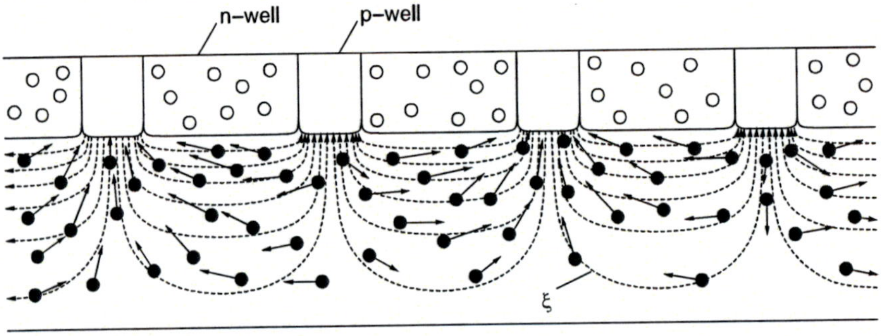

Fig. 3.17 Layout and working principle of a speed-enhanced n-well/p-sub photodiode

No metal layers are placed on top of the light-sensitive junctions, which means that no photons are reflected. The speed improvement in this case is realized by generating an electric field in the substrate. By means of this electric field, the slowly diffusing electrons are forced to drift in its direction. Because the electron concentration varies with depth, the diffusion mechanism is still active. This means that the resulting trajectory of the electrons is the sum of the diffusion speed vector, which points towards the SCR, and the drift speed vector, pointing towards one of the p-wells. Two things can therefore happen with such an electron[9]: it reaches the lower edge of the SCR and contributes to the total photocurrent or it reaches one of the p-wells and fails to contribute to the photocurrent. Electrons that are generated close to the edge of the SCR only need to diffuse over a short distance to reach its border. Therefore, it is very likely for such a carrier to reach the edge of the SCR before reaching a p-well. By contrast, electrons that are generated deeper in the substrate need to diffuse over a longer distance to reach the border of the SCR. These carriers are thus exposed to the force of the electric field for a significantly longer time. As a result, it is more probable for such an electron to reach a p-well before reaching the SCR. In conclusion, the slower electrons are removed from the photodiode current while the faster electrons are retained, which is comparable to what is achieved with a differential photodiode. The major difference between the two types of photodiode is that the magnitude of the generated photocurrent is twice as large for the speed-enhanced photodiode thanks to the absence of a reflective metal mask on top of the structure.

The electric field is generated by applying a voltage difference, V_{field}, between adjacent p-wells. Consequently, the electric potential in the substrate is no longer constant which it was before applying this voltage difference. The evolution of the electric potential in the substrate when a voltage difference V_{field} is applied between two adjacent p-wells of a speed-enhanced photodiode with a periodicity of half the photodiode diameter is shown in Fig. 3.18. If it is assumed that the potential evolves linearly in the intervals $[-1/2, -1/4]$, $[-1/4, 0]$, $[-1/2, -1/4]$ and $[-1/2, -1/4]$ for a periodicity w_p of $d/2$, the magnitude of the electric field exactly below the SCR can be estimated as follows:

$$|\xi| = \frac{2V_{field}}{w_p}. \tag{3.35}$$

For a photodiode with a diameter of 60 μm, a periodicity of 30 μm and an applied voltage difference of 1 V, the electric field has a calculated magnitude of 66.7 kV/m. Consequently, the electrons in the substrate reach a drift velocity v_{e^-} which can be quantified with (3.27) to equal 8.67 μm/ns if the electron mobility in the substrate is 1300 cm^2/Vs. Because of that, the slowly diffusing electrons in the substrate are drained very efficiently.

To shift the trade-off between the photodiode speed and responsivity, the magnitude of the electric field in the substrate can be changed; this alters the drift velocity

[9] Only the electrons within one diffusion length are considered here. Electrons that are generated deeper in the substrate are assumed to recombine.

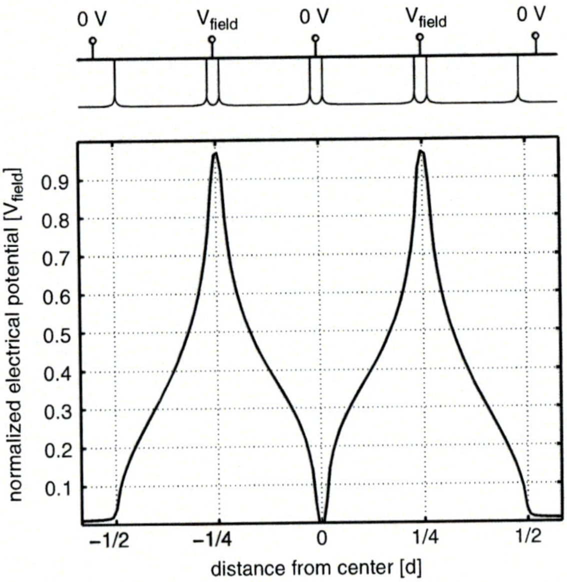

Fig. 3.18 Normalized electric potential in the substrate of a speed-enhanced photodiode with a diameter d and a periodicity of $d/2$ exactly below the SCR of the n-well/p-sub junctions; simulated with *COMSOL*

of the electrons accordingly. This can be obtained by varying V_{field} or by choosing a different periodicity. Whereas the latter method can only be applied during the layout of the photodiode, the former method can be used during operation.

3.2.3 The Photodiode Parasitics

The electrical model of a photodiode, which is shown Fig. 3.11, consists of a current source, a capacitance C_{pd} and a resistance R_{pd}. The magnitude and frequency behavior of the current source have been treated in Sect. 3.2.1. The objective of this section is to quantify the photodiode series resistance and parallel capacitance.

The series resistance is the distributed path resistance that is experienced by a generated carrier while traveling towards one of the contacts. Its value depends strongly on the layout of the structure and is therefore difficult to calculate. Nevertheless, it is common practice to neglect this resistance, which is valid as long as its value is significantly lower than the input resistance of the TIA. This can be guaranteed by providing the photodiode with sufficiently many contacts to the n-well and the p-type substrate to keep R_{pd} low enough.

The parallel capacitance is the junction capacitance of the photodiode. Its value depends on the doping concentrations of the n-type and p-type regions and the voltage across the junction, as expressed by (3.12). The total junction capacitance of a CMOS photodiode can be calculated as follows:

$$C_{bp} = A \cdot C_{bp} + P \cdot C_{sw}, \tag{3.36}$$

where A and P are the area and perimeter of the photodiode, and C_{bp} and C_{sw} are the *bottom-plate capacitance* and *sidewall capacitance* of the junction respectively. While the bottom-plate capacitance is the capacitance per unit area of an n-well/p-sub junction, the sidewall capacitance is the capacitance per unit length of an n-well/p-well junction. This is because a contact to the substrate is always established by means of a p-well, which causes the sides of the n-well to be adjacent to a higher doped p-well instead of the p-type substrate. The simulated C_{bp} and C_{sw} for four different mainstream CMOS technologies as a function of the reverse voltage are shown in Fig. 3.19. The nominal supply voltage of the technology is used as a limit for the reverse voltage which is an advantage for the technology nodes with a larger minimum linewidth and an accordingly higher nominal supply voltage. It can be noticed in Fig. 3.19a that, for a certain reverse voltage, the bottom-plate capacitance for the different technologies differs significantly which is, according to (3.12), due to different doping concentrations that are used in the technologies considered. The

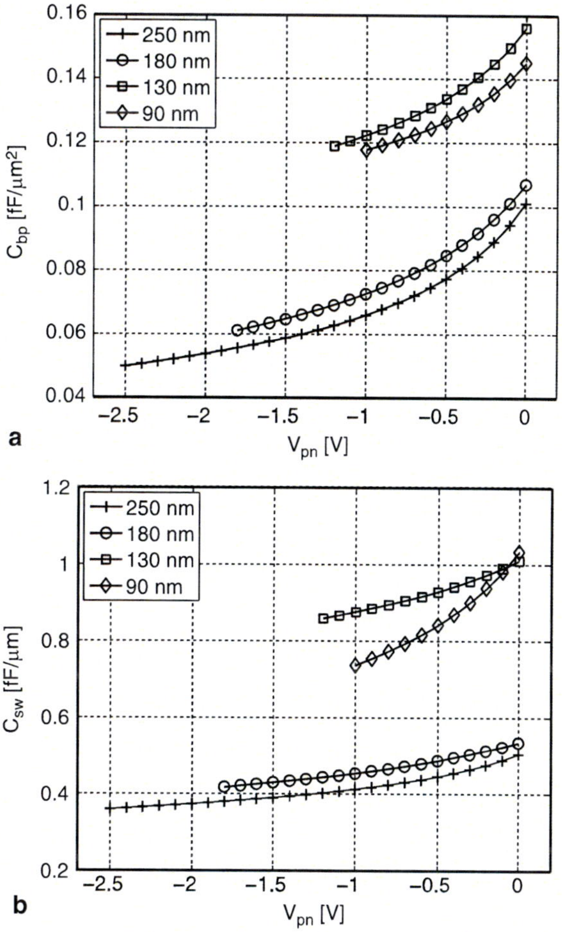

Fig. 3.19 Simulated C_{bp} and C_{sw} of an n-well/p-sub junction as a function of the reverse voltage for four mainstream CMOS technologies; the reverse voltage is limited to the nominal supply voltage of the technology: 2.5 V, 1.8 V, 1.2 V and 1 V for the 250 nm, 180 nm, 130 nm and 90 nm technology nodes, respectively. **a** C_{bp} decreases if the reverse voltage across the junction is increased. **b** C_{sw} decreases if the reverse voltage across the junction is increased; instead of showing the erroneous simulated values for the 180 nm technology node, plausible values are shown instead which have been calculated by an interpolation between the other technology nodes

Table 3.5 Simulated C_{pd} of an n-well/p-sub photodiode with a diameter of 60 µm ($A = 2830\ \mu m^2$; $P = 189\ \mu m$) for four mainstream CMOS technologies; instead of showing the erroneous simulated values for the sidewall capacitance in the 180 nm technology node, plausible values are given which have been calculated by an interpolation between the other technology nodes

	250 nm		180 nm		130 nm		90 nm	
V_{pn} [V]	−2.5	0	−1.8	0	−1.2	0	−1	0
$A \cdot C_{bp}$ [fF]	141	286	173	303	336	441	333	410
$P \cdot C_{sw}$ [fF]	67.9	95.0	74.8	101	162	190	139	195
C_{pd} [fF]	209	381	248	404	498	631	472	605

same conclusion can be drawn for the sidewall capacitance, which is represented in Fig. 3.19b. The simulated C_{sw} for the 180 nm technology node is more than 20 times smaller than the sidewall capacitance of the same junction type in the other technologies. Consequently, it is assumed that these values are incorrect. Therefore, plausible values have been assumed in this work which have been calculated by an interpolation of the same characteristic in the other technologies. These values, which are shown in Fig. 3.19b are used in the remainder of this book instead of the erroneous simulated values. According to (3.12), C_{bp} and C_{sw} should scale as $1/\sqrt{1 - V_{pn}/V_{bi}}$ if the junctions have an abrupt doping profile. However, it can be seen in Fig. 3.19 that the simulated values decrease at a slower rate which is due to the doping profile being less abrupt than assumed.

The simulated parasitic capacitance of a photodiode with a diameter of 60 µm for the four considered CMOS technologies is listed in Table 3.5. The area and perimeter of this photodiode are assumed to equal those of a circle with a diameter of 60 µm ($A = 2830\ \mu m^2$; $P = 189\ \mu m$). In reality, however, the photodiode is laid out in a fingered structure (Fig. 3.14), which causes the perimeter to be significantly higher than what is calculated above. By contrast, the area is not greatly affected by choosing a fingered photodiode layout. Again, the values for the sidewall capacitance of the n-well/p-sub junction in the 180 nm technology are assumed values that have been calculated based upon the values of the other technology nodes. From Table 3.5, it can be concluded that C_{bp} increases for a technology node with a smaller minimum linewidth. This is caused by the reverse voltage and the lower doping concentrations which are respectively higher and lower in a technology with a larger minimum linewidth.

3.2.4 *The Reflection Coefficient of CMOS*

Part of the light power that leaves the optical fiber and impinges onto an integrated photodiode is reflected. The responsivity of the integrated photodiode is affected accordingly because reflected photons cannot create electron-hole pairs. To circumvent this problem, photodiodes that are integrated in a dedicated process have an *anti-reflective coating* (ARC) deposited on top of them, which completely eliminates the reflection of light of a certain wavelength. However, it is not possible to make use of such an ARC if the photodiode is to be integrated in a standard CMOS process as no extra processing steps can be tolerated. As a result, it is of the utmost importance to quantify the amount of light which is reflected at the surface of the photodiode. This issue is examined in this section.

Before the light reaches the silicon substrate of a CMOS photodiode, it has to pass through a stack of *inter-layer dielectrics* (ILDs). These non-conducting layers are deposited on top of the substrate in order to isolate from each other the metal layer, which are used to interconnect all devices on a chip. A cross-section of this stack of ILDs is shown in Fig. 3.20 for a 130 nm CMOS technology with eight metal layers. The thicknesses and properties of the different layers have been retrieved from the technology documentation. Because different technology nodes have a different number of metal layers, the number of dielectric layers varies accordingly.

Light traveling from one medium to another is subjected to reflection and refraction at the boundary. This is exactly what happens when light traveling through the air impinges on the dielectric stack. It is assumed that the light wave impinges perpendicularly on the dielectric stack, a situation which is thoroughly described in [Ina00]. Each layer in this stack has a permittivity ε_i, a permeability μ_i and a conductivity σ_i. The intrinsic impedance for each layer is then:

$$\eta_i = \sqrt{\frac{\mu_i}{\varepsilon_i}}. \tag{3.37}$$

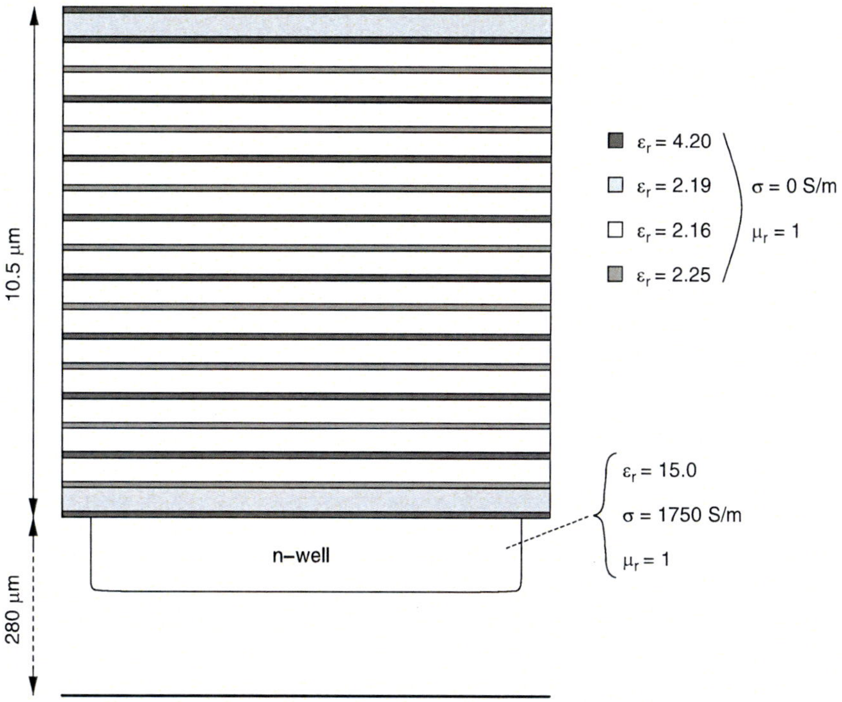

Fig. 3.20 Cross-section of a 130 nm CMOS process; a 10.5 μm thick stack of ILDs is placed on top of the 280 μm silicon substrate

Generally, the permittivity ε_i of a material is a complex number which depends on frequency. The imaginary part of ε_i is a measure of the losses due to absorption that appear when an electromagnetic wave passes through the material. For metals, which have a high conductivity, and for dielectrics, which have a zero conductivity, the complex permittivity can be written as follows[10]:

$$\varepsilon_{metal} = \varepsilon_i - j\frac{\sigma}{\omega}, \tag{3.38}$$

$$\varepsilon_{dielectric} = \varepsilon_i' - j\varepsilon_i'', \tag{3.39}$$

where ω is the angular frequency of the electromagnetic wave. Whereas ε_{metal} is always complex, $\varepsilon_{dielectric}$ is always real, except in three very narrow frequency bands where $\varepsilon_i'' > 0$ [Ina00]. This is shown in Fig. 3.21. Consequently, a conductive material always absorbs the energy of an electromagnetic wave, while a dielectric material absorbs its energy only in these three frequency bands. Electromagnetic waves having a frequency in the *ultra high frequency* (UHF) range or the microwave band are absorbed because of the resonance of dipoles in the material. By contrast, waves in the infrared and ultraviolet bands are absorbed due to ionic and electronic resonance, respectively. The real part of $\varepsilon_{dielectric}$ drops correspondingly in these three frequency bands, as shown in Fig. 3.21. As the frequency range of interest for CMOS photodiodes is between the infrared and ultraviolet region, the permittivity is assumed to be purely real for all dielectric layers on top of the silicon substrate. The relative permittivities that are given in Fig. 3.20 are valid in this frequency range. By contrast, because silicon is a conducting material, which is especially the case for the n-well, it always has a complex permittivity, as formulated in (3.38).

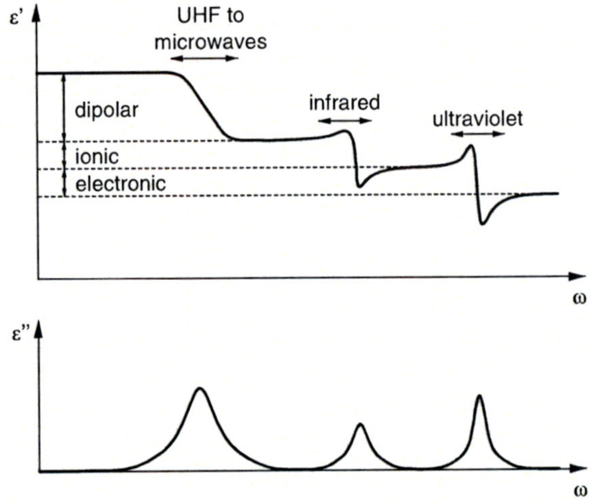

Fig. 3.21 Real and complex permittivity of a dielectric material as a function of angular frequency

[10] This difference is due to the different loss mechanisms in a metal and a dielectric.

For light traveling from a medium with intrinsic impedance η_1 to a medium with intrinsic impedance η_2, the reflection coefficient for the electric and magnetic fields is given by:

$$\Gamma_{1,2} = \frac{\eta_2 - \eta_1}{\eta_2 + \eta_1}. \quad (3.40)$$

Things get more complicated for structures with more than two layers. A portion of the light is reflected at each boundary and the other portion is refracted. The reflected part undergoes the same process on the previous boundary while the refracted part undergoes the same process on the next boundary. This goes on and on until a steady state is reached. Consequently, the total reflected wave is composed of an infinite number of components. Attention must be paid to the phase shift that is experienced by the wave when traveling through a layer with thickness d_i. Due to this phase shift, two waves can interfere constructively when they have the same phase whereas they can (partly) eliminate each other when they have different phases. This can be accounted for by generalizing (3.40) to the following:

$$\Gamma_{1,n} = \frac{\eta_{2,n} - \eta_1}{\eta_{2,n} + \eta_1}, \quad (3.41)$$

where $\Gamma_{1,n}$ is the reflection coefficient for an electromagnetic wave that impinges from a medium with intrinsic impedance η_1 onto a structure with $n-1$ other layers. In (3.41), $\eta_{2,n}$ represents the impedance of the stack of the $n-1$ other layers as experienced at the boundary between the first medium and the rest of the stack. Its value depends on all layers in the stack except for the first one:

$$\eta_{2,n} = \eta_2 \frac{1 + \Gamma_{2,n} e^{-2\gamma_2 d_2}}{1 - \Gamma_{2,n} e^{-2\gamma_2 d_2}}, \quad (3.42)$$

where γ_2 is the propagation constant in the second medium which can be calculated as follows:

$$\gamma_2 = \sqrt{j\omega\mu_2(\sigma_2 + j\omega\varepsilon_2)}. \quad (3.43)$$

In (3.42), $\Gamma_{2,n}$ is the reflection coefficient when light impinges from the second medium onto the stack consisting of the other $n-2$ layers. This can be written recursively as follows:

$$\Gamma_{i,n} = \frac{\eta_{i+1,n} - \eta_i}{\eta_{i+1,n} + \eta_i}, \quad (3.44)$$

$$\eta_{i+1,n} = \eta_{i+1} \frac{1 + \Gamma_{i+1,n} e^{-2\gamma_{i+1} d_{i+1}}}{1 - \Gamma_{i+1,n} e^{-2\gamma_{i+1} d_{i+1}}}, \quad (3.45)$$

$$\gamma_{i+1} = \sqrt{j\omega\mu_{i+1}(\sigma_{i+1} + j\omega\varepsilon_{i+1})}, \quad (3.46)$$

where $i = 1 \ldots n-1$ and $\Gamma_{n,n} = 0$, which means that there is no reflected wave after the last boundary. In this specific case, the nth layer is actually the silicon substrate.

Because there are no losses in the dielectric stack, the sum of the total reflected and refracted light power equals the incident light power. The light power transmission coefficient T, which is the fraction of the light power that penetrates the silicon substrate, can now be calculated as follows:

$$T = 1 - |\Gamma_{1,n}|^2. \quad (3.47)$$

The responsivity of the integrated photodiode is affected by this loss of light power. The *external responsivity*, R_{ext}, is defined as follows:

$$R_{ext} = T \cdot R_{int}, \quad (3.48)$$

where R_{int} is the responsivity as calculated in Sect. 3.2.1. From now on, R_{ext} is used to denote the responsivity of an integrated photodiode.

The calculated transmission coefficient (3.47) for four mainstream CMOS technologies is plotted in Fig. 3.22 as a function of wavelength. It can be noticed that the transmission of light depends strongly on the applied wavelength and the technology node used. A technology with a larger minimum linewidth typically has fewer

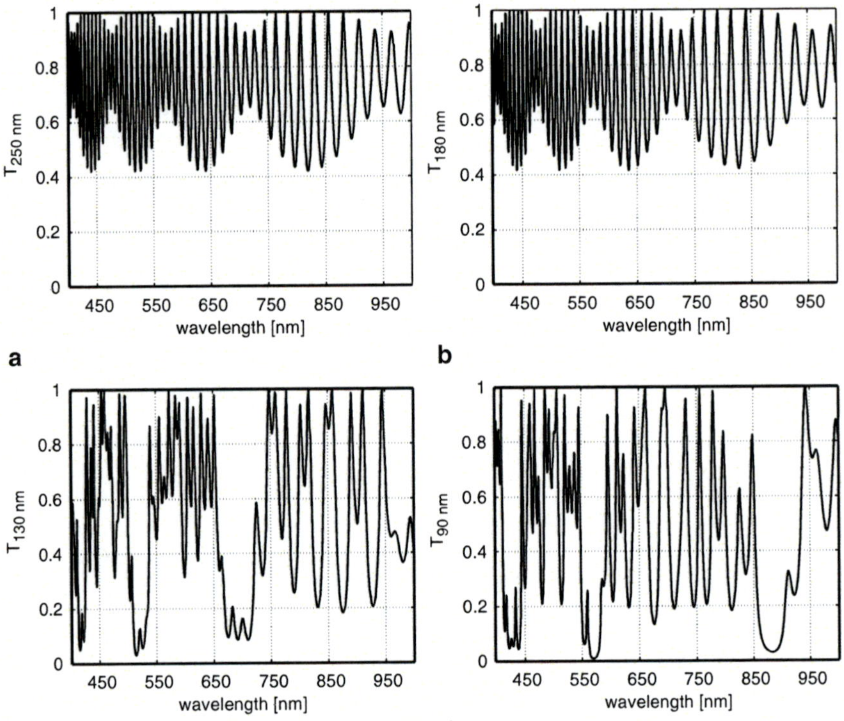

a b c d

Fig. 3.22 Calculated transmission coefficient as a function of wavelength for four mainstream CMOS technologies. **a** 250 nm CMOS. **b** 180 nm CMOS. **c** 130 nm CMOS. **d** 90 nm CMOS

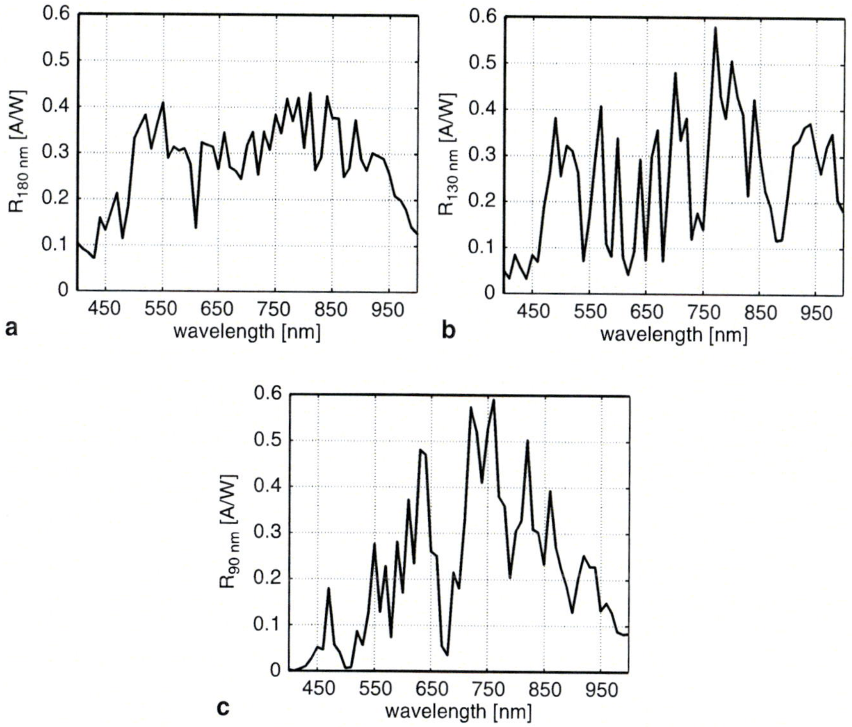

Fig. 3.23 Measured responsivity of an n-well/p-sub photodiode as a function of wavelength for three mainstream CMOS technologies. **a** 180 nm CMOS. **b** 130 nm CMOS. **c** 90 nm CMOS

interconnect metal layers and therefore also fewer dielectric layers, which results in a more regular behavior of its transmission coefficient as a function of wavelength (Fig. 3.22a, b). A technology with a smaller minimum linewidth typically has more metal layers and thus also more dielectric layers, resulting in a very unpredictable transmission coefficient (Fig. 3.22c, d). The results in Fig. 3.22 are obtained by assuming that each layer has its nominal thickness. However, it is very likely that these thicknesses change by up to 20% in production. This has a major impact on the locations of the minima and maxima of the transmission coefficient as a function of wavelength.

In order to validate the proposed model for the transmission coefficient, the responsivity of an n-well/p-sub photodiode in three mainstream CMOS technologies has been measured as a function of wavelength. The results are shown in Fig. 3.23. It can easily be seen that the responsivity of a silicon photodiode drops if the wavelength is short. In this case, a photon has a lot of energy, which means that only a few photons impinge for a certain light power, generating only a few electron-hole pairs. Moreover, the responsivity drops if the wavelength approaches the critical value of 1.11 μm above which no electron-hole pairs can be generated. The responsivity peaks for intermediate wavelengths. It can also be concluded from Fig. 3.23

that the responsivity behaves less regularly for a technology with a smaller minimum linewidth. In Sect. 3.2.1, it has been deduced that the theoretical responsivity of an n-well/p-sub photodiode is approximately 0.619 A/W for 850 nm light and 0.532 A/W for 660 nm light. It can be seen in Fig. 3.23 that the measured responsivity for a photodiode in 180 nm CMOS is only 0.378 A/W and 0.345 A/W for wavelengths of 850 nm and 660 nm, respectively. The responsivity of a photodiode in 130 nm CMOS is 0.306 A/W for 850 nm light and 0.295 A/W for 660 nm light. In 90 nm CMOS, an n-well/p-sub photodiode has a measured responsivity of 0.233 A/W and 0.250 A/W for the two wavelengths of interest.

3.3 Conclusion

The first task of an optical receiver is to convert the signal from the optical domain to the electrical domain. This is performed by the photodiode which is treated extensively in this chapter.

The working principle of a photodiode is treated first. The concepts of radiative generation and Shockley-Read-Hall recombination are shown to be the dominant physical mechanisms in a silicon photodiode. The carriers that are generated by the impinging light need to be collected in order to establish a useful photocurrent. In a photodiode this is done by the electric field that is present in the SCR. Carriers that are generated inside the SCR form a drift current while carriers that are generated in the neutral regions where no electric field is present form electron and hole diffusion currents. While the former current contribution is fast, the diffusion current is very slow. In order to obtain a photodiode with high responsivity and high speed, it is important to maximize the width of the SCR.

Next, the discussion is concentrated on photodiodes that are integrated in a standard CMOS technology. It is shown that three different junction types can be realized in this kind of process and that none of them has a wide SCR. Therefore, the majority of the photons is absorbed outside the SCR and especially in the p-type substrate due to the low absorption coefficient of silicon for light with a wavelength of 850 nm. As a result, the slow electron diffusion component originating in the p-type substrate dominates the photodiode response; this results in a conversion speed in the lower MHz range. Because of this, four techniques are discussed in order to increase the speed: using a shorter wavelength, a p+/n-well with guard, a differential photodiode layout, and the newly developed speed-enhanced photodiode. The last solution combines the speed advantage of the differential photodiode with the responsivity advantage of a normal n-well/p-sub photodiode. Finally, it is shown that the dielectric layers on top of the substrate in a CMOS process can have a detrimental effect on the photodiode responsivity. Measurements are shown to validate the performed calculations.

Chapter 4
From Current to Voltage—The Transimpedance Amplifier

After conversion from the optical domain to the electrical domain by the photodiode, the TIA converts the small photodiode current into a voltage. This building block is treated extensively in this chapter.

The most important specifications of a TIA are discussed first: transimpedance, bandwidth, noise and overload current. As it is the first genuine electrical building block in an optical receiver, these specifications determine to a great extent the performance of the complete receiver.

Thereafter, the most common TIA topologies are analyzed. Both open-loop topologies and closed-loop topologies are examined and the trade-off between the important TIA specifications is examined for all of them.

Next, the discussion is focused on the shunt-shunt feedback TIA. The different basic architectures of the voltage amplifier are treated and some considerations are made regarding the feedback network.

Finally, the design procedure and the simulation results are presented for three TIA designs.

4.1 Important Specifications

Transimpedance The main objective of a TIA is to convert the small photocurrent into a voltage. The output signal of a TIA is therefore a voltage, while the input signal is a current. The relationship between both is given by the *transimpedance* Z_{TIA}:

$$Z_{TIA} = \frac{v_{out}}{i_{in}}, \tag{4.1}$$

where v_{out} is the small-signal output voltage of the TIA and i_{in} is its small-signal input current. This is shown schematically in Fig. 4.1. When the input current is delivered by a photodiode, i_{in} is the small-signal current of the photodiode i_{pd}. Because the current at the input is transformed into a voltage at the output, Z_{TIA} has the units of Ω or dBΩ.

F. Tavernier, M. Steyaert, *High-Speed Optical Receivers with Integrated Photodiode in Nanoscale CMOS,* Analog Circuits and Signal Processing 5,
DOI 10.1007/978-1-4419-9925-2_4, © Springer Science+Business Media, LLC 2011

Fig. 4.1 Input and output signals of a TIA

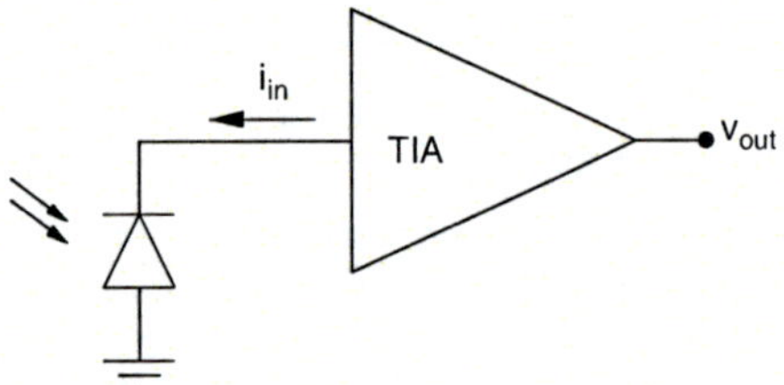

In general, v_{out} and i_{in} are phasors[1] with magnitudes and phases which depend on frequency. As a consequence, the transimpedance Z_{TIA}, which is the transfer function of a TIA, is also a frequency dependent phasor. As a result, it can be written as follows:

$$Z_{TIA}(f) = |Z_{TIA}(f)| \cdot e^{j\Phi_{Z_{TIA}}(f)}, \tag{4.2}$$

where $|Z_{TIA}(f)|$ and $\Phi_{Z_{TIA}}(f)$ are the frequency dependent magnitude and phase of the transimpedance. It can be deduced from (4.2) that the transimpedance is a complex number. From the following relationship:

$$e^{j\Phi_{Z_{TIA}}(f)} = \cos(\Phi_{Z_{TIA}}(f)) + j\sin(\Phi_{Z_{TIA}}(f)), \tag{4.3}$$

it can be concluded that the transimpedance as defined in (4.2) is real only if $\Phi_{Z_{TIA}}(f)$ equals $k \cdot 180°$ where k is a natural number. The transimpedance is then more precisely called the *transresistance*. Typically, this is the case at lower frequencies at which capacitors and inductors do not yet have an effect on the transimpedance.

The magnitude of the transimpedance should be chosen carefully. If $|Z_{TIA}(f)|$ is too low, the output signal of the TIA is small, which means that the noise of the subsequent building blocks has a significant impact on the SNR of the system. On the contrary, if $|Z_{TIA}(f)|$ is too high, v_{out} is large and it is probable that the TIA circuit leaves its linear region of operation; this introduces severe pulse-width distortion and jitter.

In theory, the magnitude and phase of the transimpedance can be arbitrary functions of frequency. In reality, however, the frequency dependency originates from poles and zeros in the circuit that are caused by capacitances or inductances. If the transimpedance has only a single pole f_p, originating from a time constant τ_p, (4.2) can be rewritten as follows:

$$Z_{TIA}(f) = Z_{TIA,DC} \cdot \frac{1}{1 + \underbrace{j2\pi f \tau_p}_{=j\frac{f}{f_p}}}, \tag{4.4}$$

where $Z_{TIA,DC}$ represents the magnitude of the transimpedance at DC. The frequency dependent magnitude and phase of the transimpedance can now be expressed more

[1] A phasor or phase vector is a representation of a sine wave with a time-invariant amplitude and phase.

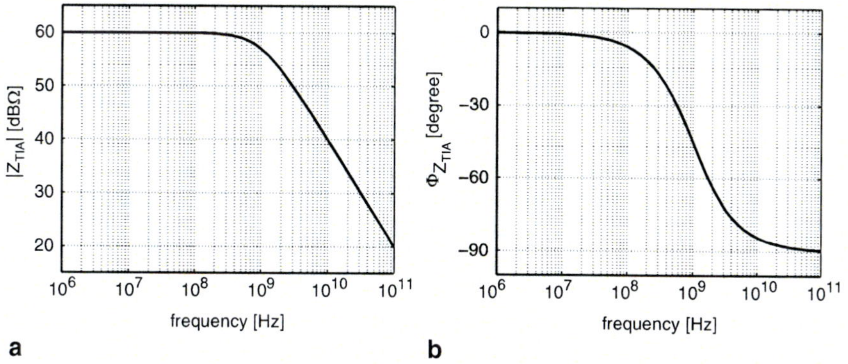

Fig. 4.2 Bode plot of the transfer function of a TIA with a transimpedance at DC of 1 kΩ = 60 dBΩ and a pole at 1 GHz. **a** Magnitude. **b** Phase

specifically:

$$|Z_{TIA}(f)| = \frac{|Z_{TIA,DC}|}{1 + \left(\frac{f}{f_p}\right)^2} \tag{4.5}$$

$$\Phi_{Z_{TIA}}(f) = \underbrace{-\arctan\frac{f}{f_p}}_{\text{radian}} = \underbrace{-\frac{180°}{\pi}\arctan\frac{f}{f_p}}_{\text{degree}}. \tag{4.6}$$

The Bode plot[2] of the transimpedance of a TIA with a single pole at 1 GHz and a transimpedance at DC of 1 kΩ = 60 dBΩ is shown in Fig. 4.2. At the pole frequency f_p, the magnitude of the transimpedance is 3 dB below the DC value as can be seen in Fig. 4.2a. Therefore, the pole frequency is also called the −3 dB frequency if the system contains only a single pole. The amplitude of the output voltage being 3 dB lower than at DC, the output power of the TIA is 3 dB, or a factor of 2, lower at the pole frequency compared to its DC value. The phase of the transimpedance experiences a phase shift of −45° at f_p, as can be seen in Fig. 4.2b where the phase is plotted in degrees. The output voltage v_{out} is delayed with respect to the input current i_{in} by an amount t_{pd} which is called the *phase delay* and can be calculated as follows:

$$t_{pd}(f) = -\frac{\Phi_{Z_{TIA}}(f)}{360°} \cdot \frac{1}{f}, \tag{4.7}$$

where $\Phi_{Z_{TIA}}$ is in degrees. For the example with a single pole at 1 GHz, the phase delay of the output voltage can be calculated to be 125 ps at the pole frequency. For frequencies $f \ll f_p$, $|\Phi_{Z_{TIA}}|$ increases linearly with frequency as can be derived

[2] A Bode plot is a graph that shows the logarithm of the magnitude and the phase of a system's transfer function versus frequency, plotted with a logarithmic frequency axis.

from (4.6). As a result, t_{pd} has a constant value at these frequencies, which is equal to 159 ps in this example. By contrast, the output voltage has a constant phase shift of $-90°$ with respect to the input current for frequencies $f \gg f_p$ so that t_{pd} evolves asymptotically to 0 s. To conclude, the phase delay between the input current and the output voltage exhibits strong frequency dependency for frequencies around and above the pole frequency which consequently introduces DDJ.

It has been shown in Chap. 2 that the -3 dB frequency, or bandwidth, of a system's transfer function determines the attainable bit rate. If data at B bit/s is to be received, the bandwidth should be $0.6 - 0.7 \cdot \mathrm{B}$ Hz as an optimal trade-off between signal distortion and integrated output noise. Together with the DC transimpedance, **the bandwidth is consequently one of the most important specifications of this** building block. Both are combined in the *transimpedance-bandwidth product ZBW* [Ing04]:

$$ZBW = |Z_{TIA,DC}| \cdot BW_{TIA}, \tag{4.8}$$

where BW_{TIA} is the -3 dB frequency of the TIA. It will be derived in Sect. 4.2 that ZBW **is fixed for a certain TIA topology, which means that the transimpedance gain** can only be increased if the bandwidth is decreased.

Noise It was shown in Chap. 2 that circuit noise reduces the sensitivity of a system. This is especially important for the TIA because its input signals can be very small due to the low responsivity of the photodiode. As in any circuit, the noise can be referred to the input where both a current noise source and a voltage noise source represent the effective circuit noise. This is shown in Fig. 4.3 for a TIA with an input impedance $Z_{in,TIA}$ and a transimpedance Z_{TIA} connected to a photodiode with a capacitance C_{pd}. The output noise voltage spectral density is then given by the following formula:

$$\begin{aligned}\overline{dv_{n,out}^2} &= \overline{di_{n,in}^2} \cdot \left| \frac{Z_{TIA}}{1 + j2\pi f Z_{in,TIA} C_{pd}} \right|^2 \\ &\quad + \overline{dv_{n,in}^2} \cdot \left| \frac{j2\pi f Z_{TIA} C_{pd}}{1 + j2\pi f Z_{in,TIA} C_{pd}} \right|^2, \end{aligned} \tag{4.9}$$

where $\overline{di_{n,in}^2}$ and $\overline{dv_{n,in}^2}$ are the input-referred current and voltage noise spectral densities of the TIA, respectively. Evidently, their values and frequency behavior are

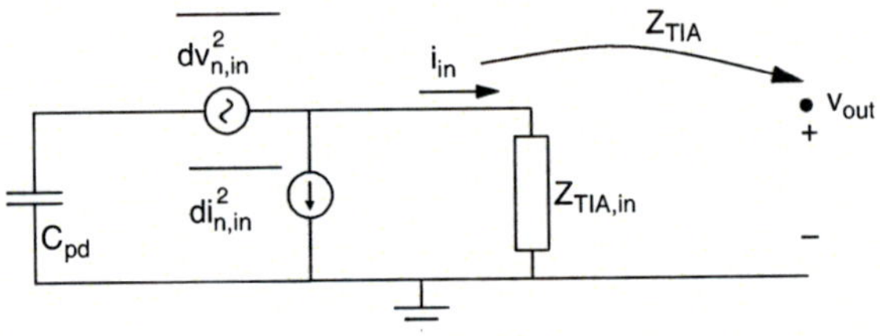

Fig. 4.3 Noise model of a TIA with an input impedance $Z_{in,TIA}$ and a transimpedance Z_{TIA} connected to a photodiode with a capacitance C_{pd}

entirely determined by the circuit topology and the specific sizing of all devices involved. It can be deduced from (4.9) that the only important noise contribution at low frequencies comes from the current noise $\overline{di_{n,in}^2}$ because the voltage noise $\overline{dv_{n,in}^2}$ is blocked thanks to the photodiode capacitance C_{pd}. By contrast, C_{pd} shorts the current noise source at very high frequencies and thereby eliminates its effect on the output noise voltage. In that case, the complete output noise voltage is determined by the input noise voltage alone. Note, however, that $\overline{di_{n,in}^2}$ is derived from $\overline{dv_{n,out,o}^2}$, the TIA output noise voltage spectral density when the input of the TIA is left open, while $\overline{dv_{n,in}^2}$ is derived from $\overline{dv_{n,out,s}^2}$, the TIA output noise voltage spectral density when the input of the TIA is short-circuited. As discussed in Sect. 2.5.2, the input noise sources of the TIA can be calculated as follows:

$$\overline{dv_{n,in}^2} = \frac{\overline{dv_{n,out,s}^2}}{|TF_V|^2} = \frac{\overline{dv_{n,out,s}^2}}{\left|\frac{Z_{TIA}}{Z_{in,TIA}}\right|^2}, \tag{4.10}$$

$$\overline{di_{n,in}^2} = \frac{\overline{dv_{n,out,o}^2}}{|TF_I|^2} = \frac{\overline{dv_{n,out,o}^2}}{|Z_{TIA}|^2}, \tag{4.11}$$

where, as expected, $TF_I = Z_{TIA}$ is the TIA transfer function for an input current and $TF_V = Z_{TIA}/Z_{in,TIA}$ is the TIA transfer function for an input voltage.

As always, the RMS noise voltage at the output can be calculated by means of the following formula:

$$v_{n,out,RMS} = \sqrt{\int_0^\infty \overline{dv_{n,out}^2}}. \tag{4.12}$$

As mentioned before, the input-referred voltage noise source does not contribute to the output noise voltage at low frequencies due to the photodiode capacitance. To simplify the calculations, it is therefore often assumed that the input-referred noise of the TIA consists only of a current noise source with the following value [Her07]:

$$i_{n,in,RMS} = \frac{v_{n,out,RMS}}{|Z_{TIA,DC}|}. \tag{4.13}$$

As this noise current source appears in parallel with the photodiode current, both values can simply be added to determine the input current of the TIA. In the end, the circuit noise determines the BER P_e if digital data is received. Its value can be calculated using the following formula:

$$P_e = Q\left(\frac{i_{ptp}}{2i_{n,in,RMS}}\right), \tag{4.14}$$

where i_{ptp} is the peak-to-peak signal current. The above formula is valid only if the noise of the photodiode can be neglected, which is typically the case due to the low responsivity of an integrated photodiode.

Overload Current While noise is important for small input signals, distortion is important for large input signals. Both reduce the obtainable BER. Although the binary nature of digital data implies that distortion can be tolerated to some extent, feedback circuits tend to behave poorly if the stages are driven into saturation. The *overload current* of the TIA is defined as the maximum input current for which the BER remains below a certain level, typically 10^{-12}. The ratio of the overload current to the minimum input current which is limited by circuit noise is usually called the *dynamic range*. Due to the low nominal supply voltage of contemporary CMOS technologies, the dynamic range of CMOS circuits in general and a TIA in particular can be too low to amplify the input signal adequately over its complete dynamic range. In this case, it can be useful to design a multiple-gain TIA which extends the effective dynamic range by reducing the transimpedance for large input signals [Che09].

As shown in the previous chapter, the responsivity of an integrated photodiode is rather low, which means that the TIA is driven into saturation only for very high optical power levels. Therefore, the overload current limit is not a major concern when using an integrated photodiode.

4.2 Comparison of TIA Topologies

4.2.1 Open-Loop Topologies

Single-Resistor TIA The basic objective of a TIA is to convert a current into a voltage. A resistor is able to do this. Consequently, the simplest TIA topology one can imagine consists of a single resistor. This is shown in Fig. 4.4, where the resistor is connected between the supply voltage and the output node, while the anode of the photodiode with a parasitic capacitance C_{pd} is tied to ground. This is the only viable configuration if an integrated n-well/p-sub photodiode is used because the p-type substrate is always connected to ground. The resistor R_{TIA} ensures that the photodiode is always reverse biased as long as the photodiode current i_{pd} obeys the following condition:

$$i_{pd} < \frac{V_{dd}}{R_{TIA}}, \tag{4.15}$$

where V_{dd} is the supply voltage.

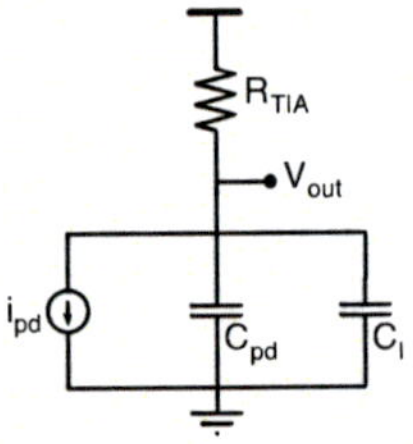

Fig. 4.4 Circuit diagram of the single-resistor TIA connected to a photodiode with a parasitic capacitance C_{pd}

The transimpedance of the single-resistor TIA can be expressed as follows:

$$Z_{TIA} = \frac{-R_{TIA}}{1 + j2\pi f R_{TIA}(C_{pd} + C_l)}, \tag{4.16}$$

where C_l represents the load capacitance of the TIA, which is typically significantly smaller than C_{pd}. The minus sign originates from the 180° phase shift between the input current and the output voltage. The magnitude of the DC transimpedance and the −3 dB bandwidth of the single-resistor TIA are, respectively:

$$|Z_{TIA,DC}| = R_{TIA}, \tag{4.17}$$

$$BW_{TIA} = \frac{1}{2\pi R_{TIA}(C_{pd} + C_l)}. \tag{4.18}$$

Note that resistor R_{TIA} defines the gain of the TIA as well as its bandwidth. As a consequence, a fundamental trade-off exists between both performance figures which is clarified by the transimpedance-bandwidth product:

$$ZBW = \frac{1}{2\pi(C_{pd} + C_l)}, \tag{4.19}$$

which is constant for a photodiode with a particular parasitic capacitance.

The single-resistor TIA has only a single noise source, namely R_{TIA}, as shown in Fig. 4.5. The voltage noise spectral density of a resistor has a pink noise as well as a white noise component. It is assumed in this work that only white noise is present[3]. The output noise voltage spectral density is then given by the following formula:

$$\begin{aligned}\overline{dv_{n,out}^2} &= \frac{\overline{dv_{n,R_{TIA}}^2}}{\left|1 + j2\pi f R_{TIA}(C_{pd} + C_l)\right|^2} \\ &= \frac{4kT R_{TIA} \cdot df}{1 + \left(2\pi f R_{TIA}(C_{pd} + C_l)\right)^2}.\end{aligned} \tag{4.20}$$

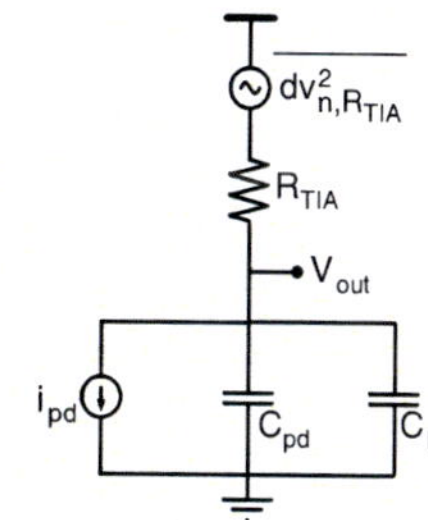

Fig. 4.5 Circuit diagram of the single-resistor TIA with the only noise source

[3] The pink noise can be neglected compared to the white noise if the bandwidth of the circuit is much higher than the frequency at which the white noise starts to dominate the pink noise.

The RMS output noise voltage can be obtained by integrating the above formula:

$$\begin{aligned} v_{n,out,RMS} &= \sqrt{\int_0^{\infty} \overline{dv_{n,out}^2}} \\ &= \sqrt{\frac{kT}{C_{pd}+C_l}}. \end{aligned} \tag{4.21}$$

The equivalent RMS input noise current can be approximated by dividing the RMS output noise voltage by the transimpedance at DC:

$$\begin{aligned} i_{n,in,RMS} &= \frac{v_{n,out,RMS}}{R_{TIA}} \\ &= \sqrt{\frac{kT}{R_{TIA}^2(C_{pd}+C_l)}}. \end{aligned} \tag{4.22}$$

From a noise point of view, it seems advantageous to increase the value of R_{TIA} because the output signal increases for a certain input signal (4.17) and the bandwidth decreases (4.18); both have a positive effect on the SNR at the output. In reality, however, the bandwidth needs to have a value of $0.6 - 0.7 \cdot \text{B}$ Hz for a bit rate of B bit/s, as mentioned before.

As an example, suppose that a bandwidth of 1 GHz is needed and that the load capacitance C_l equals 50 fF. The value of R_{TIA} can be calculated using (4.18) if C_{pd} is known. In that case, the RMS input noise current of the single-resistor TIA can be determined using (4.22). The result of this is shown in Fig. 4.6 for a range of values of C_{pd}. It can be concluded that the input-referred noise current of the single-resistor TIA increases for an increasing photodiode capacitance if the bandwidth is fixed.

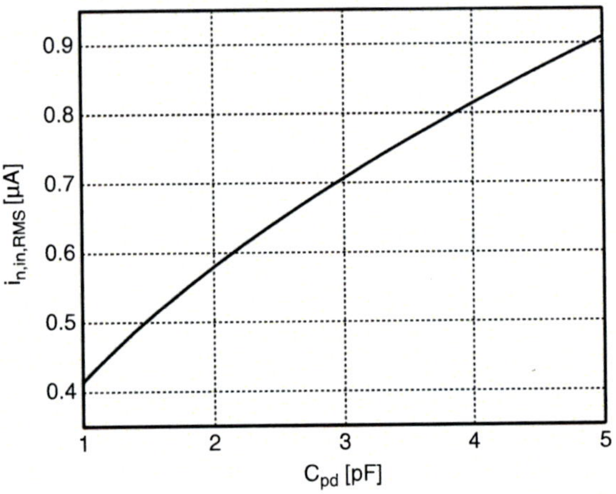

Fig. 4.6 Calculated $i_{n,in,RMS}$ as a function of C_{pd} for the single-resistor TIA with $BW_{TIA} = 1$ GHz and $C_l = 50$ fF

Finally, note that the single-resistor TIA does not consume any DC power. Power is only consumed when light impinges on the photodiode and a current flows through R_{TIA}. Clearly, this is advantageous in low-power applications.

Common-Gate TIA The problem of the single-resistor TIA is the limited transimpedance-bandwidth product which is governed only by the capacitance of the photodiode. The common-gate TIA improves this fundamental trade-off between transimpedance gain and bandwidth. The circuit diagram of a common-gate TIA connected to a photodiode with a parasitic capacitance C_{pd} is shown in Fig. 4.7.

The photodiode current i_{pd} is converted into a voltage by resistor R_{TIA}, as in the single-resistor TIA. However, instead of being determined by R_{TIA}, the input impedance of the common-gate TIA is determined by g_{m,M_1}, namely the transconductance of transistor M_1. Consequently, the bandwidth and transimpedance gain are decoupled from each other. When the output impedance of the transistors is neglected, the transimpedance of the common-gate TIA shown in Fig. 4.7 can be expressed as follows:

$$Z_{TIA} = \frac{-R_{TIA}}{\left(1 + j2\pi f \frac{C_{in}}{g_{m,M_1}}\right) \cdot (1 + j2\pi f R_{TIA} C_l)}, \tag{4.23}$$

where $C_{in} \approx C_{pd} + C_{gs,M_1}$ represents the total input capacitance. Again the minus sign denotes the 180° phase difference between the input current and the output voltage. In contrast to the transfer function of a single-resistor TIA, the common-gate TIA has a second order response. If $C_{in}/g_{m,M_1} \gg R_{TIA} C_l$, the dominant pole is located at the input node, while the non-dominant pole is at the output node. The DC transimpedance gain and the bandwidth of this TIA topology are then given respectively by:

$$|Z_{TIA,DC}| = R_{TIA}, \tag{4.24}$$

$$BW_{TIA} = \frac{g_{m,M_1}}{2\pi C_{in}}. \tag{4.25}$$

Transimpedance and bandwidth are decoupled from each other, as can be seen in the transimpedance-bandwidth product of this topology:

$$ZBW = \frac{g_{m,M_1} R_{TIA}}{2\pi C_{in}}. \tag{4.26}$$

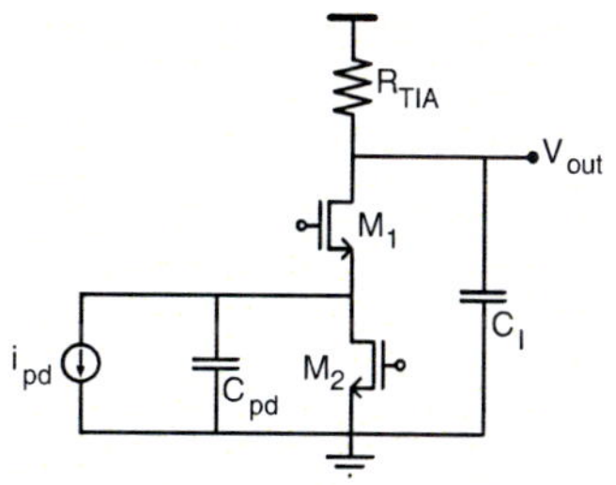

Fig. 4.7 Circuit diagram of the common-gate TIA connected to a photodiode with a parasitic capacitance C_{pd}

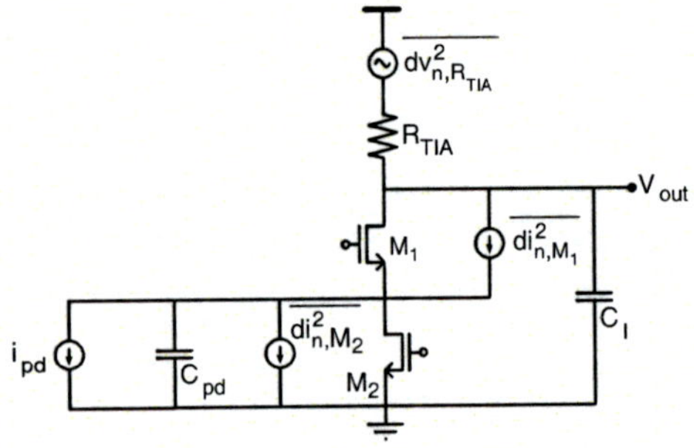

Fig. 4.8 Circuit diagram of the common-gate TIA with all noise sources added

This product can be increased by increasing R_{TIA} or g_{m,M_1} which respectively improve the transimpedance gain and the bandwidth. For a fixed gate-source voltage V_{gs,M_1}, increasing the transconductance means increasing the width of transistor M_1 and consequently raising its drain-source current I_{ds,M_1}. As this current has to flow through R_{TIA} and therefore affects the DC output voltage, the values of R_{TIA} and g_{m,M_1} cannot be chosen arbitrarily because a correct biasing condition needs to be guaranteed for the transistors in order to be able to operate correctly.

The noise behavior of the common-gate TIA is determined by three noise sources, as shown in Fig. 4.8. The noise of each transistor is represented by a current source between its drain and source instead of a voltage source at its gate, like in (2.20). As always, the relationship between the drain-source current and the gate voltage is determined by the transconductance. If only white noise is considered and if the noise of the gate, source and bulk resistances can be neglected compared to the noise of the channel resistance, the current noise spectral density of a transistor M_x depends only on its transconductance g_{m,M_x}:

$$\overline{di^2_{n,M_x}} = 4kT\gamma g_{m,M_x} \cdot df. \tag{4.27}$$

Under these assumptions, the output noise voltage spectral density of the common-gate TIA is determined by the following formula:

$$\begin{aligned}\overline{dv^2_{n,out}} = \overline{dv^2_{n,R_{TIA}}} &\cdot \left|\frac{1}{1 + j2\pi f R_{TIA} C_l}\right|^2 \\ + \overline{di^2_{n,M_1}} &\cdot \left|\frac{j2\pi f \frac{R_{TIA} C_{in}}{g_{m,M_1}}}{\left(1 + j2\pi f \frac{C_{in}}{g_{m,M_1}}\right) \cdot (1 + j2\pi f R_{TIA} C_l)}\right|^2 \\ + \overline{di^2_{n,M_2}} &\cdot \left|\frac{R_{TIA}}{\left(1 + j2\pi f \frac{C_{in}}{g_{m,M_1}}\right) \cdot (1 + j2\pi f R_{TIA} C_l)}\right|^2 .\end{aligned} \tag{4.28}$$

Due to the second order transfer function of both transistor noise components, the integrated output noise voltage results in a rather complicated expression which also depends on the location of the non-dominant pole at the output node:

$$
\begin{aligned}
v_{n,out,RMS} &= \sqrt{\int_0^{\infty} \overline{dv_{n,out}^2}} \\
&= \left[\frac{kT}{C_l} + \frac{1}{C_{in} + g_{m,M_1} R_{TIA} C_l} \right. \\
&\quad \left. \times \left(\frac{kT\gamma g_{m,M_1} R_{TIA} C_{in}}{C_l} + kT\gamma g_{m,M_1} g_{m,M_2} R_{TIA}^2\right)\right]^{\frac{1}{2}}.
\end{aligned} \tag{4.29}
$$

The first term in this formula clearly represents the noise of R_{TIA} which is integrated on the output capacitance C_l. The second term describes the noise contribution of transistors M_1 and M_2 at the output node. As before, the RMS input noise current can be approximated by dividing (4.29) by the DC transimpedance gain:

$$
\begin{aligned}
i_{n,in,RMS} &= \frac{v_{n,out,RMS}}{R_{TIA}} \\
&= \left[\frac{kT}{R_{TIA}^2 C_l} + \frac{1}{C_{in} + g_{m,M_1} R_{TIA} C_l} \right. \\
&\quad \left. \times \left(\frac{kT\gamma g_{m,M_1} C_{in}}{R_{TIA} C_l} + kT\gamma g_{m,M_1} g_{m,M_2}\right)\right]^{\frac{1}{2}}.
\end{aligned} \tag{4.30}
$$

As an example, suppose that a common-gate TIA needs to have a bandwidth of 1 GHz and that its load capacitance C_l equals 50 fF. Suppose also that the capacitance at the input is completely dominated by the photodiode capacitance ($C_{in} = C_{pd}$). The value of g_{m,M_1} can be calculated with (4.25) if C_{pd} is known. However, to calculate the RMS input noise current in (4.30), some extra information is needed. If it is assumed that the non-dominant pole has to be placed at a frequency greater than or equal to $f_{nd} = 3$ GHz[4], R_{TIA} should be at most 1.06 kΩ. The transconductance g_{m,M_2} of the biasing transistor M_2 depends on its drain-source current, I_{ds,M_2}, and its gate-source voltage, V_{gs,M_2}. For a transistor in saturation, the following relationship is valid:

$$
g_m = \frac{2I_{ds}}{V_{gs} - V_{th}}, \tag{4.31}
$$

where V_{th} is the *threshold voltage* of the transistor. From (4.30), it can be deduced that g_{m,M_2} should be minimized in order to minimize the circuit noise. Because the currents through M_1 and M_2 are the same, their transconductances are also the same if they are subjected to the same $V_{gs} - V_{th}$. From (4.30) it follows that g_{m,M_2} should be minimized in order to minimize the RMS input noise of this TIA. Therefore, it is assumed here that M_2 has a higher overdrive voltage than M_1 so that $g_{m,M_2} = g_{m,M_1}/2$. Furthermore, if it is assumed that γ equals 4/3, which is valid

[4] This criterion is imposed to ensure that the bandwidth is determined by the pole on the input node. It has nothing to do with stability because there is no feedback loop in the common-gate TIA.

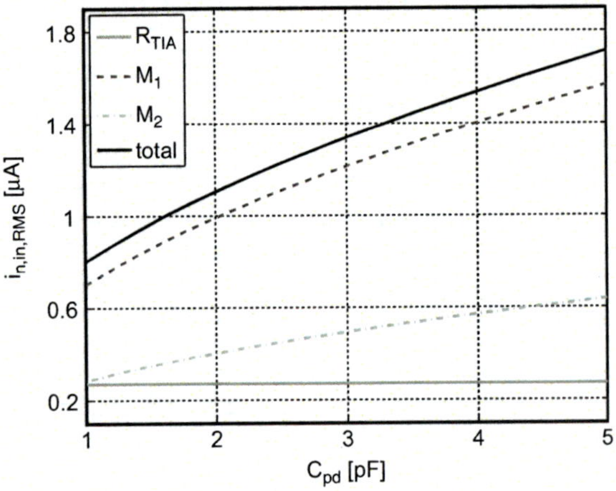

Fig. 4.9 Calculated $i_{n,in,RMS}$ as a function of C_{pd} for the common-gate TIA with $BW_{TIA} = 1$ GHz, $g_{m,M_2} = g_{m,M_1}/2$, $C_l = 50$ fF and $f_{nd} = 3$ GHz

for short-channel devices as explained in Sect. 2.5.2, the RMS input noise current can be calculated with (4.30). The result of this is shown in Fig. 4.9 for a range of values of C_{pd}. Clearly, the RMS input noise current of this TIA is determined predominantly by M_1. In contrast to the noise of R_{TIA} which does not depend on C_{pd}, both the noise of M_1 and M_2 increase with the photodiode capacitance because their transconductance needs to rise accordingly so as to maintain the same bandwidth. When compared with the results of the single-resistor TIA in Fig. 4.6, it can be noticed that the common-gate TIA produces a larger RMS input noise current. If it can be tolerated that the non-dominant pole is placed below 3 GHz, R_{TIA} can be increased, thereby reducing $i_{n,in,RMS}$. However, this has only a limited effect because the noise contribution of R_{TIA} in (4.30) is small anyhow which is also confirmed in Fig. 4.9.

By contrast with the single-resistor TIA, the common-gate TIA needs a biasing current in order to operate properly. For a photodiode capacitance of 1 pF and a bandwidth of 1 GHz, a transconductance of 6.28 mS is needed for transistor M_1. If the overdrive voltage of this transistor equals 200 mV, its drain-source current must equal 628 µA, as can be calculated with (4.31). For a supply voltage of 1.2 V, which is the nominal supply voltage of a 130 nm CMOS technology, the total power consumed by the common-gate TIA is 754 µW.

4.2.2 Closed-Loop Topologies

Regulated-Cascode TIA Feedback can be applied to the common-gate TIA to increase its performance. This leads to the regulated-cascode TIA, which is shown in Fig. 4.10. By contrast with the common-gate TIA, the gate voltage of transistor M_1 is not fixed in the regulated-cascode TIA topology. By amplifying the voltage at the source of this transistor with an inverting voltage gain, the effective transconductance of M_1 is increased by the same amount; this increases the location of the pole on

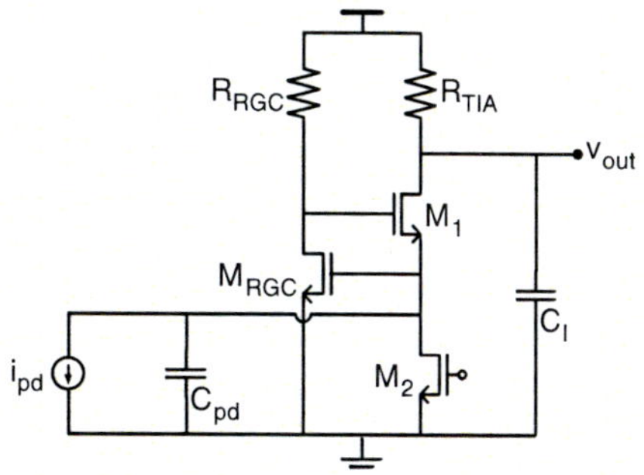

Fig. 4.10 Circuit diagram of the regulated-cascode TIA connected to a photodiode with a parasitic capacitance C_{pd}

the input node accordingly. As can be seen in Fig. 4.10, the circuit of the regulated-cascode TIA has three nodes, leading to a transfer function with three poles. Because the analysis of such a transfer function does not provide much insight, it is assumed that no pole is introduced by the extra node at the gate of transistor M_1 or that this pole is located at a very high frequency. The transimpedance of the regulated-cascode TIA is then given by:

$$Z_{TIA} = \frac{-R_{TIA}}{\left(1 + j2\pi f \frac{C_{in}}{g_{m,M_1}(1+|A_{RGC}|)}\right) \cdot (1 + j2\pi f R_{TIA} C_l)}, \tag{4.32}$$

where $|A_{RGC}| = g_{m,M_{RGC}} R_{RGC}$ is the absolute value of the voltage gain between the source and gate of M_1, or equivalently, between the gate and drain of M_{RGC}. The total input capacitance C_{in} is now determined by the photodiode capacitance and the parasitic capacitances of transistors M_1 and M_{RGC}. Taking the *Miller effect* into account, C_{in} can be calculated as follows:

$$C_{in} = C_{pd} + C_{gs,M_{RGC}} + \underbrace{(1 + |A_{RGC}|)}_{\text{Miller effect}} \cdot \left(C_{gs,M_1} + C_{gd,M_{RGC}}\right), \tag{4.33}$$

If it is assumed again that the dominant pole is located at the input, the DC transimpedance and the bandwidth of this topology are given respectively by:

$$|Z_{TIA,DC}| = R_{TIA}, \tag{4.34}$$

$$BW_{TIA} = \frac{g_{m,M_1}(1 + |A_{RGC}|)}{2\pi C_{in}}. \tag{4.35}$$

The transimpedance-bandwidth product of the regulated-cascode TIA can be expressed with the following formula:

$$ZBW = \frac{g_{m,M_1}(1 + |A_{RGC}|) R_{TIA}}{2\pi C_{in}}. \tag{4.36}$$

Compared with the common-gate TIA, the transconductance of M_1 can be decreased while the bandwidth is still maintained, or equivalently, the bandwidth can be increased for the same transconductance of M_1.

Because the regulated-cascode TIA contains a feedback loop, stability needs to be guaranteed. This can be done by ensuring that the gain in the loop is smaller than

1 when the phase has shifted by 180°. If it is assumed that the circuit contains a pole at the input and one at the output, the loop gain LG is given as follows:

$$LG = \frac{-|A_{RGC}|}{1 + j2\pi f \frac{C_{in}}{g_{m,M_1}}}. \tag{4.37}$$

Because there is only one pole in this loop gain, leading to a maximum phase shift of 90°, the regulated-cascode TIA seems to be intrinsically stable. In reality, however, an extra pole at $f_{p,RGC}$ is present at the gate of M_1 introducing an extra 90° phase shift. In order to remain stable, $f_{p,RGC}$ should satisfy the following condition:

$$f_{p,RGC} \geq 3 \cdot \frac{g_{m,M_1}|A_{RGC}|}{2\pi C_{in}} \approx 3 \cdot BW_{TIA}. \tag{4.38}$$

This condition means that the non-dominant pole in the expression for the loop gain needs to be located at a frequency which is at least 3 times higher than the frequency where the magnitude of the loop gain drops below unity. When this condition is fulfilled, the *phase margin* in the loop is at least 72° [San06], which is sufficient not to produce overshoot in the time domain. If $|A_{RGC}| \gg 1$, the condition is equivalent to stating that the pole at the gate of transistor M_1 is located at a frequency that is at least 3 times higher than the bandwidth of the regulated-cascode TIA. The value of $f_{p,RGC}$ is determined by both R_{RGC} and the capacitance at the gate of M_1. If $f_{p,RGC}$ is too low, the value of R_{RGC} can be decreased to obtain a stable system. However, this also reduces $|A_{RGC}|$ and consequently the efficacy of the feedback loop.

The noise analysis is performed based on the model depicted in Fig. 4.11. The noise voltage spectral density at the output is given by the following formula:

$$\begin{aligned}
\overline{dv^2_{n,out}} = {} & \overline{dv^2_{n,R_{TIA}}} \cdot \left| \frac{1}{1 + j2\pi f R_{TIA} C_l} \right|^2 \\
& + \overline{di^2_{n,M_1}} \cdot \left| \frac{j2\pi f \frac{R_{TIA} C_{in}}{g_{m,M_1}(1 + |A_{RGC}|)}}{\left(1 + j2\pi f \frac{C_{in}}{g_{m,M_1}(1 + |A_{RGC}|)}\right) \cdot (1 + j2\pi f R_{TIA} C_l)} \right|^2 \\
& + \overline{di^2_{n,M_2}} \cdot \left| \frac{R_{TIA}}{\left(1 + j2\pi f \frac{C_{in}}{g_{m,M_1}(1 + |A_{RGC}|)}\right) \cdot (1 + j2\pi f R_{TIA} C_l)} \right|^2 \\
& + \overline{dv^2_{n,R_{RGC}}} \cdot \left| \frac{j2\pi f \frac{R_{TIA} C_{in}}{1 + |A_{RGC}|}}{\left(1 + j2\pi f \frac{C_{in}}{g_{m,M_1}(1 + |A_{RGC}|)}\right) \cdot (1 + j2\pi f R_{TIA} C_l)} \right|^2 \\
& + \overline{di^2_{n,M_{RGC}}} \cdot \left| \frac{j2\pi f \frac{R_{TIA} R_{RGC} C_{in}}{1 + |A_{RGC}|}}{\left(1 + j2\pi f \frac{C_{in}}{g_{m,M_1}(1 + |A_{RGC}|)}\right) \cdot (1 + j2\pi f R_{TIA} C_l)} \right|^2 .
\end{aligned} \tag{4.39}$$

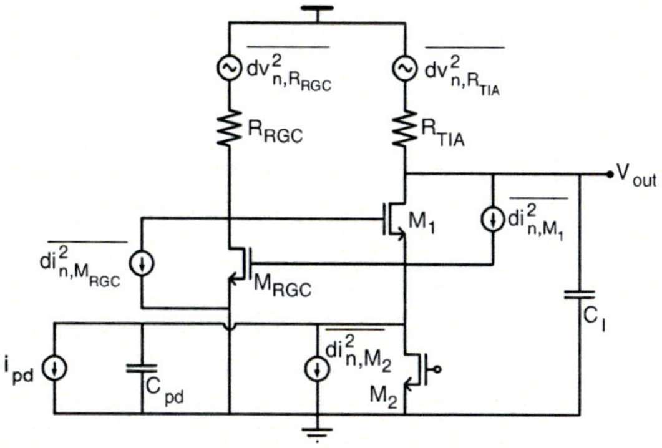

Fig. 4.11 Circuit diagram of the regulated-cascode TIA with all noise sources added

The RMS input noise current is again obtained by dividing the RMS output noise voltage by the DC transimpedance gain of the regulated-cascode TIA:

$$\begin{aligned} i_{n,in,RMS} &= \frac{\sqrt{\int_0^\infty \overline{dv^2_{n,out}}}}{R_{TIA}} \\ &= \left[\frac{kT}{R^2_{TIA} C_l} + \frac{1}{C_{in} + (1 + |A_{RGC}|) g_{m,M_1} R_{TIA} C_l} \right. \\ &\quad \times \left(\frac{kT\gamma g_{m,M_1} C_{in}}{R_{TIA} C_l} + (1 + |A_{RGC}|) kT\gamma g_{m,M_1} g_{m,M_2} \right. \\ &\quad \left. \left. + \frac{kT g^2_{m,M_1} R_{RGC} C_{in}}{R_{TIA} C_l} + \frac{kT\gamma g^2_{m,M_1} g_{m,M_{RGC}} R^2_{RGC} C_{in}}{R_{TIA} C_l} \right) \right]^{\frac{1}{2}} \end{aligned} \tag{4.40}$$

Suppose, as an example, that a regulated-cascode TIA needs to have a bandwidth of 1 GHz and that its load capacitance C_l equals 50 fF, as in the previous example. Suppose also that the capacitance at the input is again dominated completely by the photodiode capacitance ($C_{in} = C_{pd}$). If $|A_{RGC}| = 5$, which is a reasonable assumption in a contemporary CMOS process[5], the value of g_{m,M_1} can be calculated using (4.35) if C_{pd} is known. It is again assumed that the non-dominant pole at the output has to be placed at a frequency of at least $f_{nd} = 3$ GHz. As a result, the value of R_{TIA} should again be no larger than 1.06 kΩ. It is also supposed again that $g_{m,M_2} = g_{m,M_1}/2$ and that γ equals 4/3. The transconductance of M_{RGC} can, at first sight, be chosen arbitrarily. However, it is determined by $|A_{RGC}|$, which is chosen equal to 5, and R_{RGC}, which defines the stability, as predicted by (4.38). For the moment, $g_{m,M_{RGC}}$ is chosen equal to g_{m,M_1}. Using (4.40), the RMS input noise current of the regulated-cascode TIA can be calculated. The result of this is

[5] The gain $|A_{RGC}| = g_{m,M_{RGC}} R_{RGC}$ can be made arbitrarily high by increasing R_{RGC}. However, this is not true when the finite output resistance of a transistor is taken into account. This finite output resistance leads to a finite intrinsic transistor gain and therefore also to a limited $|A_{RGC}|$.

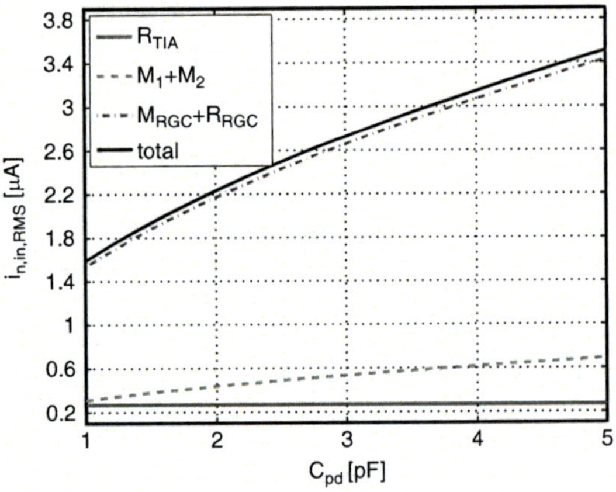

Fig. 4.12 Calculated $i_{n,in,RMS}$ as a function of C_{pd} for the regulated-cascode TIA with $BW_{TIA} = 1$ GHz, $g_{m,M_2} = g_{m,M_1}/2$, $g_{m,M_{RGC}} = g_{m,M_1}$, $|A_{RGC}| = 5$, $C_l = 50$ fF and $f_{nd} = 3$ GHz

shown in Fig. 4.12 for a range of values of C_{pd}. It can be concluded that $i_{n,in,RMS}$ is completely dominated by the noise of M_{RGC} and R_{RGC}. Moreover, the noise of M_1 and M_2 is much smaller than in the common-gate TIA because their transconductance can be $|A_{RGC}|$ times smaller. On the contrary, the noise contribution of R_{TIA} is the same as in the previous topology because the non-dominant pole is located at the same frequency. Comparing the results of the single-resistor TIA in Fig. 4.6 and the common-gate TIA in Fig. 4.9, note that the regulated-cascode TIA produces a larger RMS input noise current.

For a photodiode capacitance of 1 pF, a bandwidth of 1 GHz, and $|A_{RGC}| = 5$, the transconductance of M_1 needs to be 1.05 mS, which is $1 + |A_{RGC}|$ times smaller than in a common-gate TIA. If the overdrive voltage of M_1 equals 200 mV, its drain-source current is 105 µA (4.31). The same current flows through M_{RGC} if the same overdrive voltage is used and if $g_{m,M_{RGC}} = g_{m,M_1}$. For a supply voltage of 1.2 V, the nominal supply voltage of a 130 nm CMOS technology, the power consumption of the regulated-cascode TIA is 251 µW, which is lower than that of the common-gate TIA.

Two conclusions can be drawn from the previous discussion:

1. The noise of the regulated-cascode TIA is higher than that of the common-gate TIA under the chosen assumptions;
2. The power consumption of the regulated-cascode TIA is lower than that of the common-gate TIA with the same assumptions.

The question could be posed whether the noise of the regulated-cascode TIA would still exceed the noise of a common-gate TIA if it could consume an equal amount of power. Obviously, the extra power should be consumed by the amplifier M_{RGC}-R_{RGC} and not by the input branch because the bandwidth and consequently g_{m,M_1} are not required to change in this example. Moreover, it has been concluded from Fig. 4.12 that the noise of the regulated-cascode TIA is dominated by the noise of M_{RGC} and R_{RGC}. If the current through both these devices is increased, $g_{m,M_{RGC}}$ rises while R_{RGC} should be decreased accordingly so as to maintain the same $|A_{RGC}|$. The

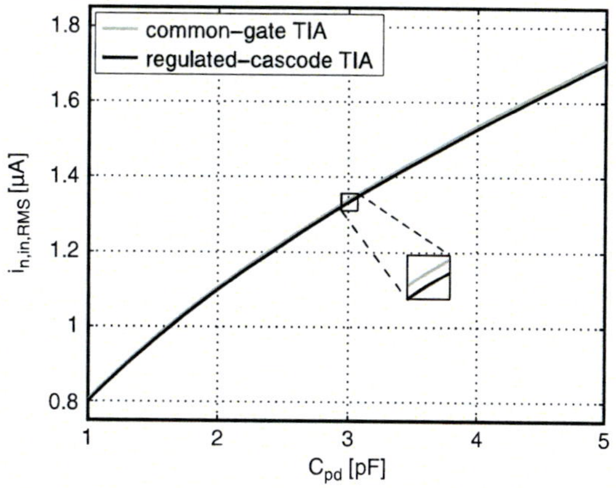

Fig. 4.13 Calculated $i_{n,in,RMS}$ as a function of C_{pd} for the regulated-cascode TIA with $BW_{TIA} = 1$ GHz, $g_{m,M_2} = g_{m,M_1}/2$, $g_{m,M_{RGC}} = |A_{RGC}|g_{m,M_1}$, $|A_{RGC}| = 5$, $C_l = 50$ fF and $f_{nd} = 3$ GHz compared with the result for the common-gate TIA in Fig. 4.9

following relationship should be valid in order to equalize the power consumption of the common-gate TIA and the regulated-cascode TIA:

$$g_{m,M_{RGC}} = |A_{RGC}|g_{m,M_1}, \tag{4.41}$$

which means that the current through transistor M_{RGC} could be increased by a factor $|A_{RGC}|$. Consequently, R_{RGC} should be decreased with the same amount to maintain the same $|A_{RGC}|$ as before. This results in a decreased RMS input noise current, as shown in Fig. 4.13, where it is compared with the RMS input noise current of the common-gate TIA of Fig. 4.9. From this, it can be concluded that the common-gate TIA and the regulated-cascode TIA generate the same input-referred noise when they are allowed to consume the same amount of power.

Shunt-Shunt Feedback TIA Yet another possibility to convert a current into a voltage is shown in Fig. 4.14. A negative feedback network senses the voltage at the output and returns a proportional current to the input. Therefore, this circuit is said to be in a voltage–current feedback configuration or, equivalently, in a shunt-shunt feedback configuration[6]. The conversion from the voltage at the output to a current

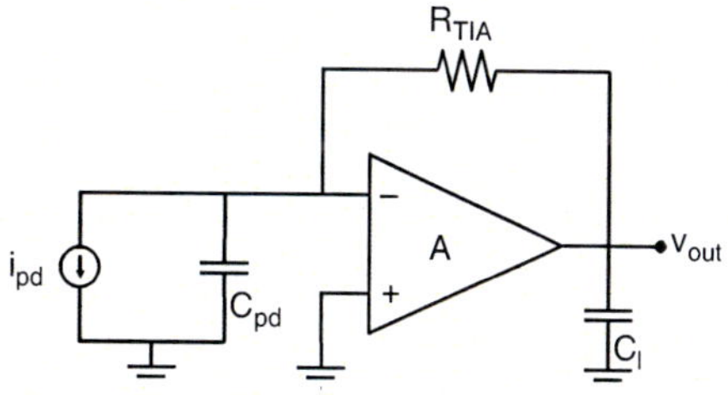

Fig. 4.14 Circuit diagram of the shunt-shunt feedback TIA connected to a photodiode with a parasitic capacitance C_{pd}

[6] A voltage is sensed by a high-impedance network whigh is placed in parallel with the voltage (shunt). The same is true for steering a current. Therefore, voltage-current feedback can also be called shunt-shunt feedback.

at the input is performed by R_{TIA}. The transimpedance of the shunt-shunt feedback TIA is given by the following formula:

$$Z_{TIA} = \frac{A R_{TIA} - R_{out,A}}{1 + A + j2\pi f \left(R_{TIA} C_{pd} + R_{out,A} \left(C_{in} + C_l\right)\right) - 4\pi^2 f^2 R_{TIA} R_{out,A} C_{in} C_l}, \tag{4.42}$$

where $R_{out,A}$ represents the output resistance of voltage amplifier A and C_l is the load capacitance. Ideally, the voltage amplifier should have infinite voltage gain A and zero output resistance. In reality, it can be assumed that A is much larger than 1 and that $R_{out,A}$ is much smaller than R_{TIA}. In that case, the transimpedance from (4.42) can be approximated with the following formula:

$$Z_{TIA} \approx \frac{R_{TIA}}{\left(1 + j2\pi f \frac{R_{TIA} C_{in}}{A}\right) \cdot \left(1 + j2\pi f R_{out,A} C_l\right)}. \tag{4.43}$$

The input capacitance C_{in} of this TIA topology comprises the large photodiode capacitance as well as the input capacitance of the voltage amplifier. As before, it is assumed that the dominant pole is located at the input node which is a reasonable assumption because of the large photodiode capacitance. In that case, the DC transimpedance and the bandwidth of the shunt-shunt feedback TIA are given respectively by:

$$|Z_{TIA,DC}| = R_{TIA}, \tag{4.44}$$

$$BW_{TIA} = \frac{A}{2\pi R_{TIA} C_{in}}. \tag{4.45}$$

By contrast with the common-gate TIA and the regulated-cascode TIA, the DC transimpedance and the bandwidth are not decoupled from each other in the shunt-shunt feedback TIA. As with the single-resistor TIA, they are both determined by R_{TIA}. However, compared to the single-resistor TIA, the shunt-shunt feedback TIA has an additional degree of freedom, namely the voltage gain A, that allows the transimpedance-bandwidth product of this TIA to be significantly higher than that of the single-resistor TIA. This is clearly visible in the following expression:

$$ZBW = \frac{A}{2\pi C_{in}}. \tag{4.46}$$

The presence of a feedback loop in the shunt-shunt feedback TIA requires that its stability is verified and guaranteed in order not to introduce overshoot or oscillations in the time domain. If it can be assumed again that $R_{out,A} \ll R_{TIA}$, then the loop gain of this topology is given by the following formula:

$$LG = \frac{-A}{\left(1 + j2\pi f R_{TIA} C_{in}\right) \cdot \left(1 + j2\pi f R_{out,A} C_l\right)}. \tag{4.47}$$

The minus sign in (4.47) originates from the fact that the feedback is negative, at least at DC. In order to be stable, the non-dominant pole of (4.47) should be located at a frequency that is at least at 3 times higher than the gain-bandwidth product of the

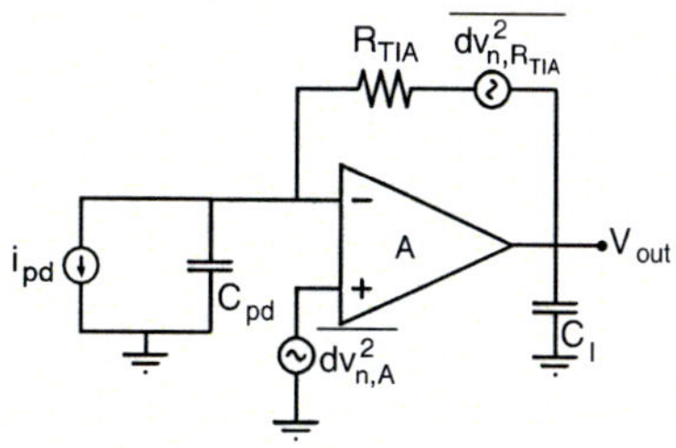

Fig. 4.15 Circuit diagram of the shunt-shunt feedback TIA with its two noise sources

loop gain [San06]. If the dominant pole is assumed to be at the input node because of the large photodiode capacitance, the non-dominant pole f_{nd} should fulfill the following condition:

$$f_{nd} = \frac{1}{2\pi R_{out,A} C_l} \geq 3 \cdot \underbrace{\frac{A}{2\pi R_{TIA} C_{in}}}_{BW_{TIA}} . \tag{4.48}$$

This means that the non-dominant pole of the loop gain, which is also the non-dominant pole of the transimpedance as given in (4.42), should be located at a frequency that is at least 3 times higher than the bandwidth of the shunt-shunt feedback TIA.

The noise analysis of the shunt-shunt feedback TIA is performed based on Fig. 4.15. The noise of R_{TIA} is represented by its voltage noise spectral density while the noise of the voltage amplifier depends on its actual circuit implementation. However, the latter can always be referred to its input, as described more thoroughly in Sect. 2.5.2. In general, this should be done by a voltage noise source as well as by a current noise source in order not to depend on the output resistance of the preceding circuit. However, if the noise is referred to the positive input which is connected to a fixed voltage in the shunt-shunt feedback TIA, the noise current can be omitted because it is shorted. This is the reason why it is not shown in Fig. 4.15. The output noise voltage spectral density is given by:

$$\begin{aligned} \overline{dv^2_{n,out}} = \overline{dv^2_{n,R_{TIA}}} \cdot & \left| \frac{1 + j2\pi f \frac{R_{out,A} C_{in}}{A}}{\left(1 + j2\pi f \frac{R_{TIA} C_{in}}{A}\right) \cdot \left(1 + j2\pi f R_{out,A} C_l\right)} \right|^2 \\ + \overline{dv^2_{n,A}} \cdot & \left| \frac{1 + j2\pi f R_{TIA} C_{in}}{\left(1 + j2\pi f \frac{R_{TIA} C_{in}}{A}\right) \cdot \left(1 + j2\pi f R_{out,A} C_l\right)} \right|^2 , \end{aligned} \tag{4.49}$$

where it is again assumed that $A \gg 1$ and $R_{out,A} \ll R_{TIA}$. In order to be able to make some initial calculations, $\overline{dv^2_{n,A}}$, the input-referred voltage noise spectral density of the voltage amplifier, is assumed to be approximately equal to the voltage noise spectral density of its input transistor M_1[7]. In that case, the RMS output noise voltage can be calculated by means of the following formula:

$$\begin{aligned} v_{n,out,RMS} &= \sqrt{\int_0^\infty \overline{dv_{n,out}^2}} \\ &= \sqrt{\frac{kT}{\left(R_{TIA}^2 C_{in}^2 - A^2 R_{out,A}^2 C_l^2\right) g_{m,M_1} R_{out,A} C_l}} \cdot \\ &\quad \big[A g_{m,M_1} R_{TIA}^2 R_{out,A} C_{in} C_l + g_{m,M_1} R_{TIA} R_{out,A}^2 C_{in}^2 \\ &\quad - A^2 g_{m,M_1} R_{TIA} R_{out,A}^2 C_l^2 - A g_{m,M_1} R_{out,A}^3 C_{in} C_l \\ &\quad + \gamma A^2 R_{TIA}^2 C_{in}^2 + \gamma A R_{TIA} R_{out,A} C_{in} C_l \\ &\quad - \gamma A^3 R_{TIA} R_{out,A} C_{in} C_l - \gamma A^2 R_{out,A}^2 C_l^2\big]^{\frac{1}{2}} . \end{aligned} \tag{4.50}$$

where g_{m,M_1} **is the transconductance of the input transistor of the voltage amplifier.** Again, the input-referred RMS noise current can be calculated by dividing (4.50) by the transimpedance gain at DC:

$$\begin{aligned} i_{n,in,RMS} &= \frac{v_{n,out,RMS}}{R_{TIA}} \\ &= \sqrt{\frac{kT}{\left(R_{TIA}^2 C_{in}^2 - A^2 R_{out,A}^2 C_l^2\right) g_{m,M_1} R_{TIA}^2 R_{out,A} C_l}} \cdot \\ &\quad \big[A g_{m,M_1} R_{TIA}^2 R_{out,A} C_{in} C_l + g_{m,M_1} R_{TIA} R_{out,A}^2 C_{in}^2 \\ &\quad - A^2 g_{m,M_1} R_{TIA} R_{out,A}^2 C_l^2 - A g_{m,M_1} R_{out,A}^3 C_{in} C_l \\ &\quad + \gamma A^2 R_{TIA}^2 C_{in}^2 + \gamma A R_{TIA} R_{out,A} C_{in} C_l \\ &\quad - \gamma A^3 R_{TIA} R_{out,A} C_{in} C_l - \gamma A^2 R_{out,A}^2 C_l^2\big]^{\frac{1}{2}} . \end{aligned} \tag{4.51}$$

Suppose that a shunt-shunt feedback TIA needs to have a bandwidth of 1 GHz and that its load capacitance C_l equals 50 fF, as before. In addition, assume that the input capacitance is completely dominated by the photodiode capacitance ($C_{in} = C_{pd}$). The **voltage amplifier is assumed to deliver a fixed gain of 60 dB** $= 1000$. The value of the feedback resistor R_{TIA} is then completely determined by (4.45) if the photodiode capacitance is known. For a bandwidth of 1 GHz, the stability condition requires that the non-dominant pole, which is determined by $R_{out,A}$ and C_l is at a frequency of at least 3 GHz. Therefore, the value of $R_{out,A}$ is chosen equal to 1.06 kΩ. Finally, in order to be able to calculate the RMS input noise current of (4.51) numerically, the value of g_{m,M_1} should be known. By contrast with the common-gate TIA and

[7] **This assumption underestimates the noise of the voltage amplifier. However, if the input transistor of the voltage amplifier provides adequate gain to suppress the noise of the other components, the** noise is only slightly underestimated.

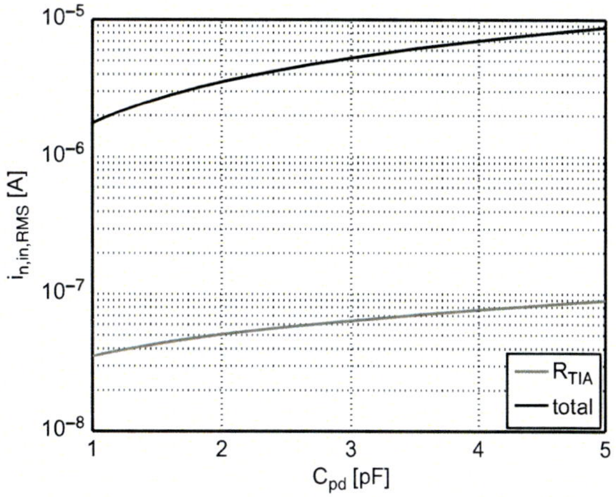

Fig. 4.16 Calculated $i_{n,in,RMS}$ as a function of C_{pd} for the shunt-shunt feedback TIA with $BW_{TIA} = 1$ GHz, $A = 60$ dB, $g_{m,M_1} = 1$ mS, $C_l = 50$ fF and $f_{nd} = 3$ GHz

the regulated-cascode TIA, this transconductance does not determine the bandwidth and can therefore be chosen freely. If $g_{m,M_1} = 1$ mS, the RMS input noise current is plotted in Fig. 4.16 as a function of the photodiode capacitance. It can be concluded easily from this figure that the noise of R_{TIA} can always be neglected as the RMS input noise current of the TIA in this example is determined solely by the amplifier. Compared with the results for the single-resistor TIA (Fig. 4.6), the common-gate TIA (Fig. 4.9) and the regulated-cascode TIA (Fig. 4.13), the input-referred noise of the shunt-shunt feedback TIA is higher and increases more rapidly as a function of C_{pd}.

The power consumption of the shunt-shunt feedback TIA depends only on the circuit design of the voltage amplifier which is unknown for the moment. However, in order to minimize its noise, it is common practice to steer the major portion of the current through its first stage and consequently through its input transistor M_1. The total current consumption of the voltage amplifier is therefore approximately equal to the current through M_1. For a transconductance of 1 mS and an overdrive voltage of 200 mV, a current of 100 μA is consumed by the shunt-shunt feedback TIA, which results in a power consumption of 120 μW if the supply voltage equals 1.2 V. Two important remarks must be made:

1. The power consumption of the common-gate TIA and the regulated-cascode TIA for a photodiode capacitance of 1 pF equal 756 μW for the same bandwidth as the shunt-shunt feedback TIA;
2. The power consumption of the common-gate TIA and the regulated-cascode TIA increase linearly as a function of the photodiode capacitance whereas the shunt-shunt feedback TIA consumes a fixed amount of power.

To make a fair comparison between these topologies, the shunt-shunt feedback TIA should be allowed to consume the same power as the other topologies. Consequently, the transconductance of the input transistor of the voltage amplifier is assumed to be

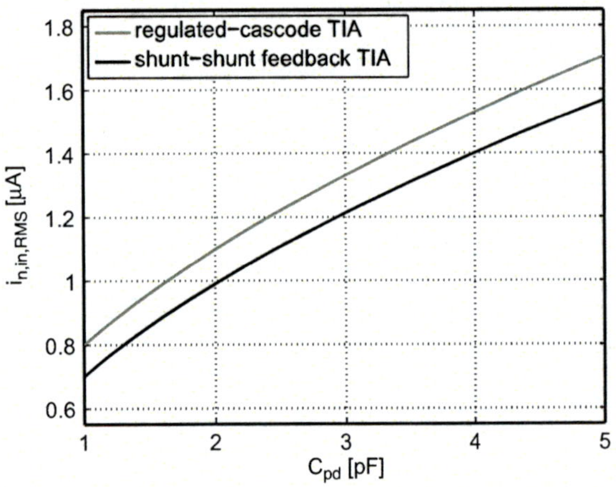

Fig. 4.17 Calculated $i_{n,in,RMS}$ as a function of C_{pd} for the shunt-shunt feedback TIA with $BW_{TIA} = 1$ GHz, $A = 60$ dB, $C_l = 50$ fF and $f_{nd} = 3$ GHz compared with the result for the regulated-cascode TIA with the same power consumption

determined by the following formula:

$$g_{m,M_1} = 2\pi BW_{TIA} C_{pd}. \tag{4.52}$$

This condition results in a larger current through the input transistor of the voltage amplifier, which also increases as a function of the photodiode capacitance. Consequently, the RMS input noise current of the shunt-shunt feedback TIA decreases significantly, as shown in Fig. 4.17 where it is also compared with the results of the regulated cascode TIA consuming the same amount of power. It can be concluded that, for the same power consumption, the shunt-shunt feedback TIA has superior noise performance compared with the regulated-cascode TIA.

From the discussion above it might be concluded that the size of the input transistor of the voltage amplifier can be chosen arbitrarily without affecting the bandwidth of the shunt-shunt feedback TIA. However, this is only true if the input capacitance of the voltage amplifier can be neglected with respect to the photodiode capacitance. For larger input devices and taking into account the Miller effect, the following formula should be used to calculate the total input capacitance of the TIA:

$$C_{in} = C_{pd} + C_{gs,M_1} + (1 + A_1)C_{gd,M_1}, \tag{4.53}$$

where A_1 is the gain of the first inverting amplifying stage and C_{gs,M_1} and C_{gd,M_1} represent the gate-source and gate-drain capacitance of the input transistor, respectively, which both depend linearly on the transistor width. Therefore, for a certain transistor length, (4.53) can be simplified by introducing parameter α:

$$C_{in} = C_{pd} + (1 + \alpha)C_{gs,M_1}. \tag{4.54}$$

This parameter α depends on the ratio of both parasitic capacitances for a certain transistor length and on the gain of the first amplifying stage. In the saturation region, the gate-source capacitance can be calculated as follows [San06]:

$$C_{gs} = \frac{2}{3} W L C_{ox}, \tag{4.55}$$

where C_{ox} is the gate oxide capacitance per unit area. The transconductance of a transistor in saturation can be related to its gate-source capacitance by considering (4.31) and (4.55):

$$g_m = \overbrace{\frac{\mu C_{ox}}{2} \frac{W}{L} \left(V_{gs} - V_{th}\right)^2}^{I_{ds}} \frac{2}{V_{gs} - V_{th}} \\ = \mu \frac{3}{2} \frac{C_{gs}}{L^2} \left(V_{gs} - V_{th}\right), \tag{4.56}$$

where μ represents the mobility of electrons in the case of an nMOS or holes when a pMOS is used. If C_{in} and g_{m,M_1} in (4.51) are eliminated by means of (4.55) and (4.56), respectively, the dependency of the RMS input noise current of a shunt-shunt feedback TIA with respect to C_{gs,M_1} is revealed. This is shown in Fig. 4.18 for a shunt-shunt feedback TIA with an nMOS input transistor with an electron mobility of 600 cm^2/Vs, connected to a photodiode with a capacitance of 1 pF. It is assumed that $\alpha = 1$ and $V_{gs} - V_{th} = 200$ mV. The other parameters are the same as before. It can be observed in this Fig. 4.18 that the RMS input noise current of the shunt-shunt feedback TIA shows a clear minimum if C_{gs,M_1} is around $0.5 \cdot C_{pd}$. The noise increases for smaller values of C_{gs,M_1} because the RMS input noise is then determined mainly by the channel noise of the input transistor, which rises for a smaller device with a correspondingly smaller transconductance. The noise increases also for larger values of C_{gs,M_1} because C_{in} is then determined completely by the input transistor. Consequently, R_{TIA} should decrease to maintain the bandwidth of 1 GHz (4.45). The RMS input noise current is then dominated by the noise of the feedback resistor.

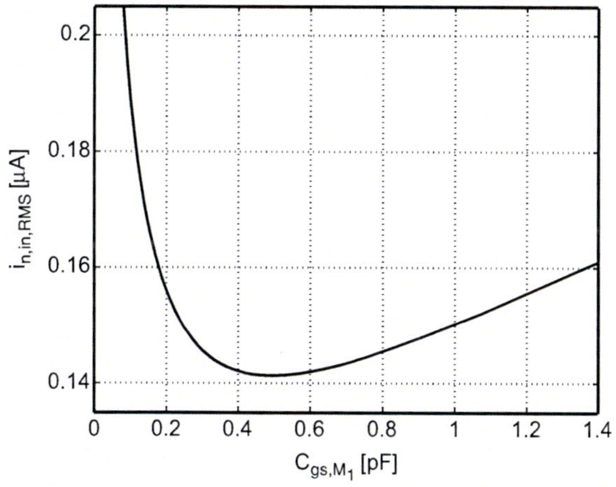

Fig. 4.18 Calculated $i_{n,in,RMS}$ as a function of C_{gs,M_1} for the shunt-shunt feedback TIA with $BW_{TIA} = 1$ GHz, $C_{pd} = 1$ pF, $\mu = 600$ cm^2/Vs, $\alpha = 1$, $V_{gs} - V_{th} = 200$ mV, $A = 60$ dB, $C_l = 50$ fF and $f_{nd} = 3$ GHz

The following noise optimizations have been proposed in the open literature:

- $C_{gs,M_1} = C_{pd}$ [Cha91]
By contrast with the derivation above, this result has been obtained by neglecting the Miller effect. Moreover, this optimum has been derived without considering the dependency of the feedback resistance on the input capacitance.
- $C_{gs,M_1} = 0.5 - 1 \cdot C_{pd}$ [Ing04]
The dependency of the feedback resistance on the input capacitance has been taken into account to calculate this optimum. However, it has been assumed again that the Miller effect can be neglected.
- $C_{gs,M_1} \approx 0.23 \cdot C_{pd}$ [Ler05]
The dependency of the feedback resistance on the input capacitance as well as the Miller effect have been considered in this optimum. The reason why it differs from the optimum in Fig. 4.18 is the different voltage gain that has been assumed for the voltage amplifier. In [Ler05] A equals 20 dB while in the discussion above it equals 60 dB. This larger voltage gain suppresses the noise of feedback resistor because its resistance can be larger for the same bandwidth. Consequently, the noise of the voltage amplifier is always dominant and the noise optimum shifts to larger values for C_{gs,M_1}.

4.3 Design Considerations of a Shunt-Shunt Feedback TIA

As can be concluded from the previous section, the shunt-shunt feedback TIA is able to combine high transimpedance gain, high bandwidth and good low-noise performance. By contrast with the other TIA topologies, it is feasible to design a differential version of the shunt-shunt feedback TIA. It is shown in Sect. 4.4 that this property is a major advantage when using a differential photodiode. For these reasons, the shunt-shunt feedback TIA is used throughout this work.

4.3.1 *Design of the Voltage Amplifier*

Until now, the voltage amplifier in the shunt-shunt feedback TIA as shown in Fig. 4.14 is considered as a black-box providing a voltage gain A over an infinite bandwidth, having an output impedance $R_{out,A}$, and an infinite input impedance. It is shown in this section how this voltage amplifier can be realized.

Common-Source Amplifier The basic voltage amplifier that can be realized in CMOS technology is the common-source amplifier. The implementation with an nMOS as the input transistor is shown in Fig. 4.19. It is also possible to use a pMOS as the input device. In that case, the load resistor R_l is placed between the output

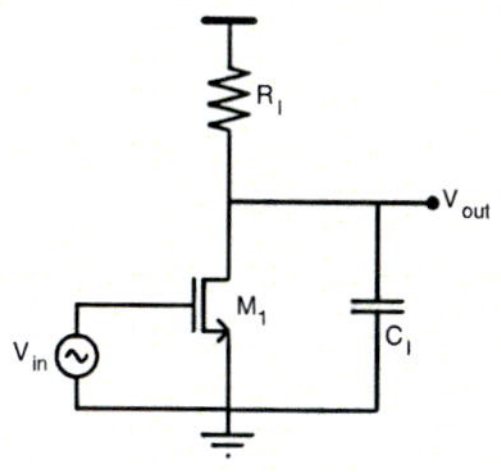

Fig. 4.19 Circuit diagram of a common-source amplifier

node and the ground instead of between the supply and the output node in the nMOS variant. The voltage gain, A_{cs}, of this common-source amplifier is defined by:

$$A_{cs} = \frac{-g_{m,M_1} R_{out,cs} \left(1 - \frac{j2\pi f C_{gd,M_1}}{g_{m,M_1}}\right)}{1 + j2\pi f R_{out,cs} \left(C_l + C_{gd,M_1}\right)}, \tag{4.57}$$

where g_{m,M_1} is the transconductance of transistor M_1, C_{gd,M_1} is its gate-drain capacitance, C_l is the load capacitance and $R_{out,cs}$ is the output resistance. For the circuit of Fig. 4.19, the output resistance is the parallel combination of the load resistance R_l and r_{ds,M_1}, which is the small-signal output resistance of transistor M_1:

$$R_{out,cs} = \frac{r_{ds,M_1} R_l}{r_{ds,M_1} + R_l}. \tag{4.58}$$

The common-source amplifier is an inverting amplifier, as indicated by the minus sign in (4.57). Both the voltage gain at DC and the bandwidth of the common-source amplifier are determined by $R_{out,cs}$:

$$\left|A_{cs,DC}\right| = g_{m,M_1} R_{out,cs}, \tag{4.59}$$

$$BW_{cs} = \frac{1}{2\pi R_{out,cs} \left(C_l + C_{gd,M_1}\right)}. \tag{4.60}$$

The DC gain of the common-source amplifier in (4.59) can be increased by increasing the value of $R_{out,cs}$. However, it should be noted that the output resistance of a transistor, r_{ds}, cannot be chosen freely as it depends on the transconductance of the transistor. Together they define the *intrinsic gain* of a transistor:

$$A_i = g_m r_{ds}, \tag{4.61}$$

which is the maximum gain that can be achieved in the 1-stage common-source amplifier shown in Fig. 4.19. The value of A_i depends on the technology used and the overdrive voltage. This is shown in Fig. 4.20b for a minimum-length transistor in a 130 nm CMOS technology. The highest intrinsic gain is obtained at the lower overdrive voltages (below 100 mV) while it decreases if the overdrive voltage is increased. It can also be concluded from Fig. 4.20b that an nMOS provides a higher

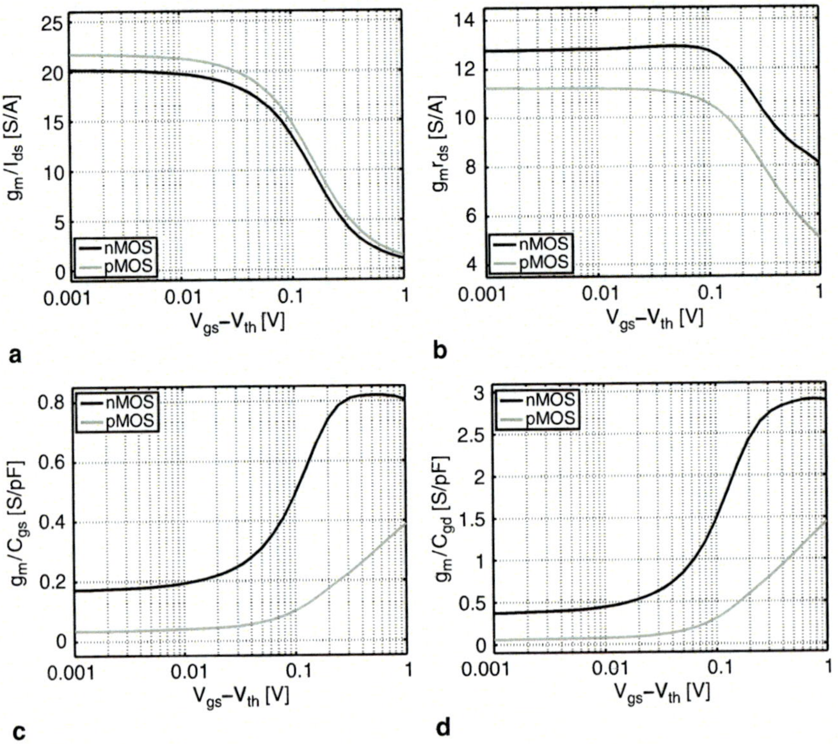

Fig. 4.20 Simulated properties of a transistor in saturation in a 130 nm CMOS technology as a function of the overdrive voltage; the transistor length is 120 nm, which is the minimum length in this technology and the width is 10 μm. **a** g_m/I_{ds}. **b** $g_m r_{ds}$. **c** g_m/C_{gs}. **d** g_m/C_{gd}

intrinsic gain than a pMOS in this technology. Moreover, the intrinsic gain can be increased if a transistor with a non-minimum length is used.

The bandwidth of the common-source amplifier in (4.60) depends on the value of $R_{out,cs}$ and the total output capacitance $C_l + C_{gd,M_1}$. It can be increased by decreasing the output resistance at the expense of reduced DC gain. It can also be increased by lowering the total output capacitance, which is ultimately limited by C_{gd,M_1} if no load capacitance is attached. For a transistor in saturation, this gate-drain capacitance originates from the overlap between the gate and the drain which depends only on the layout of the transistor. Therefore, its value does not depend on the overdrive voltage as long as the transistor stays in the saturation region ($V_{ds} \geq V_{gs} - V_{th}$)[8].

The positive zero in (4.57) originates from the gate-drain capacitance C_{gd,M_1} which provides a feedforward path from the input node to the output node. The location of

[8] For a transistor in the linear region ($V_{ds} < V_{gs} - V_{th}$) the gate-drain capacitance increases because half of the gate-channel capacitance is then referred to the drain.

this positive zero is located at a very high frequency:

$$f_{pz,cs} = \frac{g_{m,M_1}}{2\pi C_{gd,M_1}}, \tag{4.62}$$

which, for a transistor in saturation, is always higher than f_t, the *transistor unity current gain frequency* [San06]:

$$f_t = \frac{g_m}{2\pi C_{gs}}. \tag{4.63}$$

The ratios g_m/C_{gs} and g_m/C_{gd} for a minimum-length transistor in saturation in a 130 nm CMOS technology are plotted as a function of the overdrive voltage in Fig. 4.20c, d respectively. Both ratios increase significantly if the overdrive voltage is increased. This is because C_{gs} and C_{gd} do not change with the overdrive voltage as long as the transistor remains in saturation, whereas g_m does increase with an increasing overdrive voltage. The location of the positive zero in (4.62) or the transistor's f_t in (4.63) can therefore be increased by increasing the overdrive voltage.

The input impedance of the common-source amplifier in Fig. 4.19 is determined by C_{gs,M_1} and C_{gd,M_1}. Because the latter is connected to the drain, where the small-signal voltage is an inverted and amplified version of the small-signal voltage at the gate, the Miller effect should be taken into account when the input impedance of the common-source amplifier is considered:

$$\begin{aligned} Z_{in,cs} &= \frac{1}{j2\pi f\left(C_{gs,M_1} + (1 - A_{cs})\, C_{gd,M_1}\right)} \\ &= \frac{1}{j2\pi f\left(C_{gs,M_1} + \left(1 + \left|A_{cs,DC}\right|\right) C_{gd,M_1}\right)} \\ &\quad \times \frac{1 + j2\pi f R_{out,cs}\left(C_l + C_{gd,M_1}\right)}{1 + j2\pi f \frac{R_{out,cs}\left(C_{gs,M_1} C_{gd,M_1} + C_{gs,M_1} C_l + C_{gd,M_1} C_l\right)}{C_{gs,M_1} + \left(1 + \left|A_{cs,DC}\right|\right) C_{gd,M_1}}}. \end{aligned} \tag{4.64}$$

Because the voltage amplifier has a frequency dependent gain, the Miller effect also depends on frequency. The input impedance is purely capacitive at low frequencies. However, from the frequency at which the voltage gain over C_{gd,M_1} starts to decrease, namely at BW_{cs}, the input impedance becomes resistive. It becomes capacitive again for even higher frequencies where the magnitude of the voltage gain of the common-source amplifier is smaller than 1. While the input capacitance is relatively large at frequencies below the bandwidth of the common-source amplifier, it is relatively low at very high frequencies:

$$C_{in,cs} = \begin{cases} C_{gs,M_1} + \left(1 + \left|A_{cs,DC}\right|\right) C_{gd,M_1} & f < BW_{cs} \\ C_{gs,M_1} + \frac{C_{gd,M_1} C_l}{C_{gd,M_1} + C_l} & f \gg BW_{cs}. \end{cases} \tag{4.65}$$

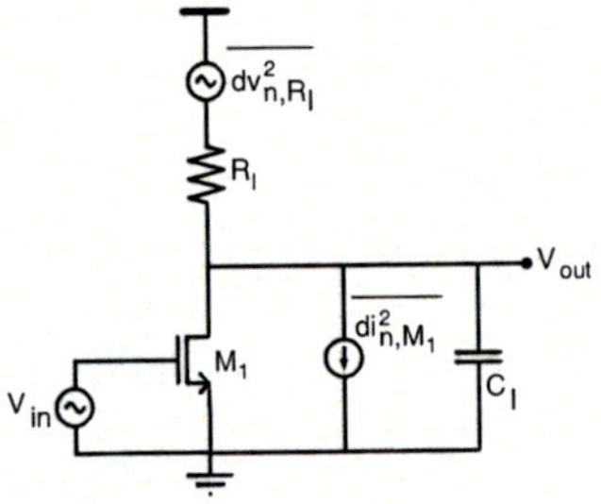

Fig. 4.21 Circuit diagram of the common-source amplifier with the two noise sources

The noise of the common-source amplifier is composed of two noise sources, as shown in Fig. 4.21. The output noise voltage spectral density is given by the following formula:

$$\overline{dv^2_{n,out,cs}} = \overline{dv^2_{n,R_l}} \cdot \left| \frac{1}{1 + j2\pi f R_{out,cs} \left(C_l + C_{gd,M_1}\right)} \right|^2 + \overline{di^2_{n,M_1}} \cdot \left| \frac{R_{out,cs}}{1 + j2\pi f R_{out,cs} \left(C_l + C_{gd,M_1}\right)} \right|^2 . \tag{4.66}$$

In order to determine the noise behavior of the shunt-shunt feedback TIA in Fig. 4.15 with a common-source amplifier, the output noise voltage spectral density should be referred to the input:

$$\begin{aligned} \overline{dv^2_{n,in,cs}} &= \frac{\overline{dv^2_{n,out}}}{|A_{cs}|^2} \\ &\approx \frac{\overline{dv^2_{n,R_l}}}{g^2_{m,M_1} R^2_{out,cs}} + \frac{\overline{di^2_{n,M_1}}}{g^2_{m,M_1}} \\ &\approx \left(\frac{4kT R_l}{g^2_{m,M_1} R^2_{out,cs}} + \frac{4kT\gamma}{g_{m,M_1}} \right) \cdot df, \end{aligned} \tag{4.67}$$

where the approximation indicates the fact that the positive zero which appears in the expression for A_{cs} is neglected for simplification.

As an example, suppose that a transconductance of 1 mS is needed for transistor M_1 and that its overdrive voltage is chosen equal to 200 mV. In addition, suppose that a 130 nm CMOS process is used and that a transistor with a minimum length is selected. The results of Fig. 4.20 are valid in this case. It follows from Fig. 4.20 that, if M_1 is an nMOS, the drain-source current equals 132 μA. If a DC voltage at the output of 0.6 V, half of the nominal supply voltage of the technology, is desired, the value of R_l can be calculated to be 4.55 kΩ. The intrinsic gain of M_1 is 11.6, as can be seen in Fig. 4.20b. This results in a small-signal output resistance of 11.6 kΩ, which leads to an $R_{out,cs}$ of 3.27 kΩ, as can be verified with (4.58). The voltage gain at DC of this common-source amplifier is thus equal to 3.27 = 10.3 dB. With

(4.67), the input noise voltage spectral density results in 5.40 $\mathrm{nV}/\sqrt{\mathrm{Hz}}$, of which 4.70 $\mathrm{nV}/\sqrt{\mathrm{Hz}}$ comes from the channel of transistor M_1 and only 2.65 $\mathrm{nV}/\sqrt{\mathrm{Hz}}$ originates in the load resistor R_l[9]. Furthermore, it follows from Fig. 4.20c, d that the gate-source capacitance and the gate drain capacitance of M_1 equal 1.36 fF and 0.413 fF, respectively. With (4.60) and a load capacitance of 5 fF, this leads to a bandwidth of 8.99 GHz. In this case, the positive zero is located at an impressively high 385 GHz (4.62). For frequencies below the bandwidth, the input capacitance of this common-source amplifier can be calculated using (4.65), which yields 3.12 fF.

Double Common-Source Amplifier Instead of steering the current of the input transistor of the common-source amplifier passively through a load resistor, this current can also be reused by a second transistor. This principle leads to the double common-source amplifier which is shown in Fig. 4.22. The voltage gain of this amplifier is given by the following formula:

$$A_{dcs} = \frac{-g_{m,M_{1+2}} R_{out,dcs} \left(1 - \frac{j2\pi f C_{gd,M_{1+2}}}{g_{m,M_{1+2}}}\right)}{1 + j2\pi f R_{out,dcs} \left(C_l + C_{gd,M_{1+2}}\right)}, \tag{4.68}$$

where $g_{m,M_{1+2}} = g_{m,M_1} + g_{m,M_2}$ and $C_{gd,M_{1+2}} = C_{gd,M_1} + C_{gd,M_2}$. The output resistance of the double common-source amplifier, $R_{out,dcs}$, is now determined by the output resistance of the two transistors:

$$R_{out,dcs} = \frac{r_{ds,M_1} r_{ds,M_2}}{r_{ds,M_1} + r_{ds,M_2}}. \tag{4.69}$$

The gain at low frequencies and the bandwidth of this amplifier are given by the following equations:

$$\left|A_{dcs,DC}\right| = g_{m,M_{1+2}} R_{out,dcs}, \tag{4.70}$$

$$BW_{dcs} = \frac{1}{2\pi R_{out,dcs} \left(C_l + C_{gd,M_{1+2}}\right)}. \tag{4.71}$$

For the same output resistance as a common-source amplifier, the gain of the double common-source amplifier is higher, while its bandwidth is only slightly lower. This

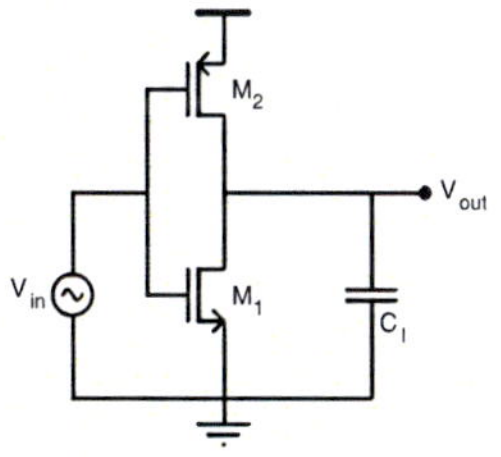

Fig. 4.22 Circuit diagram of the double common-source amplifier

[9] This validates the assumption that was made in Sect. 4.2 about the noise of the voltage amplifier in the shunt-shunt feedback TIA being dominated by the noise of the first transistor.

leads to a gain-bandwidth product which is higher for the double common-source amplifier than for the common-source amplifier. A positive zero can again be noticed in (4.68). The frequency at which it appears is given by:

$$f_{pz,dcs} = \frac{g_{m,M_{1+2}}}{2\pi C_{gd,M_{1+2}}}, \tag{4.72}$$

which is again very high.

The major difference between the common-source amplifier and the double common-source amplifier can be noticed at its input node where two transistor gates instead of one need to be charged and discharged. One of these transistors is a pMOS which has larger parasitic capacitances. Similar to the common-source amplifier in (4.64), the input impedance can be analyzed as follows:

$$\begin{aligned} Z_{in,dcs} &= \frac{1}{j2\pi f\left(C_{gs,M_{1+2}} + (1 - A_{dcs})\,C_{gd,M_{1+2}}\right)} \\ &= \frac{1}{j2\pi f\left(C_{gs,M_{1+2}} + \left(1 + \left|A_{dcs,DC}\right|\right) C_{gd,M_{1+2}}\right)} \\ &\quad \times \frac{1 + j2\pi f R_{out,dcs}\left(C_l + C_{gd,M_{1+2}}\right)}{1 + j2\pi f \frac{R_{out,dcs}\left(C_{gs,M_{1+2}}C_{gd,M_{1+2}} + C_{gs,M_{1+2}}C_l + C_{gd,M_{1+2}}C_l\right)}{C_{gs,M_{1+2}} + \left(1+\left|A_{dcs,DC}\right|\right)C_{gd,M_{1+2}}}} \end{aligned} \tag{4.73}$$

where $C_{gs,M_{1+2}} = C_{gs,M_1} + C_{gs,M_2}$. Again, the evolution of the input impedance can be divided into three regions: a capacitive region with a high input capacitance for frequencies up to BW_{dcs}, a resistive region, and a second capacitive region with a lower input capacitance for very high frequencies:

$$C_{in,dcs} = \begin{cases} C_{gs,M_{1+2}} + \left(1 + \left|A_{dcs,DC}\right|\right) C_{gd,M_{1+2}} & f < BW_{dcs} \\ C_{gs,M_{1+2}} + \dfrac{C_{gd,M_{1+2}} C_l}{C_{gd,M_{1+2}} + C_l} & f \gg BW_{dcs}. \end{cases} \tag{4.74}$$

The noise of the double common-source amplifier is produced only by transistors, as shown in Fig. 4.23. Because both noise sources experience the same transfer function

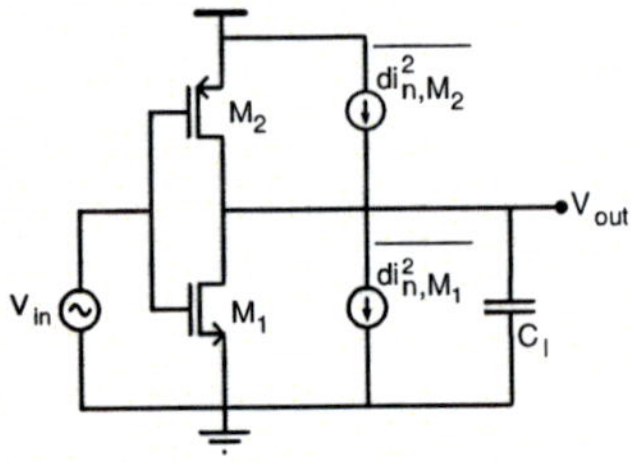

Fig. 4.23 Circuit diagram of the double common-source amplifier with its two noise sources

to the output, the output noise voltage spectral density of the double common-source amplifier is given by:

$$\overline{dv_{n,out,dcs}^2} = \left(\overline{di_{n,M_1}^2} + \overline{di_{n,M_2}^2}\right) \cdot \left| \frac{R_{out,dcs}}{1 + j2\pi f R_{out,dcs}\left(C_l + C_{gd,M_{1+2}}\right)} \right|^2. \tag{4.75}$$

If the positive zero in (4.68) is neglected, the input noise voltage spectral density can be described as follows:

$$\begin{aligned} \overline{dv_{n,in,dcs}^2} &= \frac{\overline{dv_{n,out,dcs}^2}}{|A_{dcs}|^2} \\ &\approx \frac{\overline{di_{n,M_1}^2} + \overline{di_{n,M_2}^2}}{g_{m,M_{1+2}}^2} \\ &\approx \frac{4kT\gamma}{g_{m,M_{1+2}}} \cdot df. \end{aligned} \tag{4.76}$$

As an example, suppose again that a transconductance of 1 mS is needed for transistor M_1 and that its overdrive voltage is chosen equal to 200 mV. From Fig. 4.20a it follows that its drain-source current equals 132 µA if the design is realized in a 130 nm CMOS process. The same current flows through transistor M_2 which is a pMOS. It can be derived from Fig. 4.20a that the transconductance of M_2 is slightly larger, namely 1.15 mS, if the same overdrive voltage is assumed. The intrinsic gain of the nMOS is 11.6, as before, whereas that of the pMOS is only 9.30. This leads to output resistances of M_1 and M_2 of 11.6 kΩ and 8.09 kΩ, respectively. With (4.69), the value of $R_{out,dcs}$ is 4.77 kΩ. The voltage gain of this double common-source amplifier then equals 10.3 = 20.3 dB as revealed by (4.70). With (4.76), the input noise voltage spectral density results in 3.20 nV/$\sqrt{\text{Hz}}$, which is significantly lower than what is obtained with a common-source amplifier with the same current consumption. Moreover, it follows from Fig. 4.20c, d that the gate-source capacitance and the gate drain capacitance of M_1 equal 1.36 fF and 0.413 fF, respectively, whereas the gate-source capacitance and the gate drain capacitance of M_2 equal 6.61 fF and 1.95 fF, respectively. With (4.71) and a load capacitance of 5 fF, the bandwidth of this double common-source amplifier is 4.37 GHz, which is less than half the bandwidth of the common-source amplifier with the same current consumption and load capacitance. The positive zero is now located at 145 GHz. The input capacitance for frequencies below the bandwidth can be calculated using (4.74); this gives 34.7 fF. This is more than 10 times larger than in the case of a common-source amplifier. The performance of the double common-source amplifier is summarized in Table 4.1, where it is compared to a common-source amplifier with the same power consumption.

Cascade of Amplifier Stages The voltage gain of the common-source amplifier, and even that of the double common-source amplifier, is too low to be used effectively in a shunt-shunt feedback TIA, especially when integrated in a recent CMOS technology

Table 4.1 Overview of the performance of a common-source amplifier (CS) and a double common-source amplifier (DCS) in a 130 nm CMOS process with $g_{m,M_1} = 1$ mS, an overdrive voltage of 200 mV and a load capacitance of 5 fF

	CS	DCS
gain	3.27	10.3
bandwidth	8.99 GHz	4.37 GHz
f_{pz}	385 GHz	145 GHz
C_{in}	3.12 fF	34.7 fF
input noise	5.40 nV/$\sqrt{\text{Hz}}$	3.20 nV/$\sqrt{\text{Hz}}$

with its accompanying low intrinsic gain. Two techniques can be applied to increase the voltage gain of a 1-stage amplifier:

1. Cascoding: the gain increases and the bandwidth decreases accordingly; the power consumption is the same as that of a conventional 1-stage amplifier;
2. Cascading: the gain increases and the bandwidth decreases only slightly; the power consumption is higher than that of a conventional 1-stage amplifier.

The first technique is not useful if the voltage amplifier is to be used in a shunt-shunt feedback TIA because it delivers its gain over a narrow bandwidth. The second technique, namely the cascading of 1-stage amplifiers, is preferred because it combines high gain with high bandwidth.

The block diagram of an n-stage amplifier is shown in Fig. 4.24. It is composed of n 1-stage amplifiers that can be implemented as common-source amplifiers, double common-source amplifiers or other 1-stage amplifier topologies. The DC voltage gain of each stage is represented by $A_{i,DC}$ and the finite bandwidth of every stage is modeled by a resistor R_i and a capacitor C_i, as shown in Fig. 4.24.

The voltage gain of such an n-stage amplifier, $A_{cascade}$, is given by the following formula:

$$A_{cascade} = \frac{A_{1,DC}}{1 + j2\pi f R_1 C_1} \cdot \frac{A_{2,DC}}{1 + j2\pi f R_2 C_2} \cdot \ldots \cdot \frac{A_{n,DC}}{1 + j2\pi f R_n C_n}. \qquad (4.77)$$

If every stage has an identical circuit implementation, the 1-stage DC voltage gains are the same: $A_{1,DC} = A_{2,DC} = \ldots = A_{n,DC} = A_{1\text{-}stage,DC}$. If, in addition, all the stages have the same size and thus conduct the same current, all resistors R_i

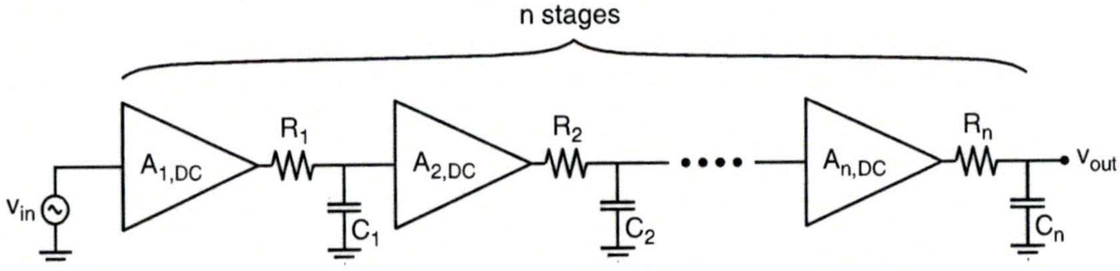

Fig. 4.24 Block diagram of an n-stage amplifier, composed of n 1-stage amplifiers

and capacitors C_i have the same values[10]: $R_1 = R_2 = \ldots = R_n = R_{1\text{-}stage}$ and $C_1 = C_2 = \ldots = C_n = C_{1\text{-}stage}$. In that case, the voltage gain of the cascade of n stages can be simplified as follows:

$$A_{cascade} = \left(\frac{A_{1\text{-}stage,DC}}{1 + j2\pi f R_{1\text{-}stage} C_{1\text{-}stage}} \right)^n . \tag{4.78}$$

Instead of a 1-stage amplifier which only has a single pole that is intrinsically the same as its −3 dB bandwidth, the voltage gain of the cascade of n 1-stage amplifiers has n coinciding poles. This results in a −3 dB bandwidth which is lower than the frequencies of each of the poles. The DC voltage gain and the −3 dB bandwidth of the n-stage amplifier are respectively given by:

$$\left|A_{cascade,DC}\right| = \left|A_{1\text{-}stage,DC}\right|^n , \tag{4.79}$$

$$BW_{cascade} = \frac{\sqrt{\sqrt[n]{2} - 1}}{2\pi R_{1\text{-}stage} C_{1\text{-}stage}} = \sqrt{\sqrt[n]{2} - 1} \cdot BW_{1\text{-}stage} . \tag{4.80}$$

If the gain of each 1-stage amplifier is significantly higher than 1, the DC voltage gain of the cascade increases very rapidly with the number of cascaded stages while the bandwidth of the cascade decreases only slowly. This is best visualized by considering the *gain-bandwidth product*, GBW, of the n-stage amplifier:

$$\begin{aligned} GBW_{cascade} &= \left|A_{1\text{-}stage,DC}\right|^n \cdot \sqrt{\sqrt[n]{2} - 1} \cdot BW_{1\text{-}stage} \\ &= \left|A_{1\text{-}stage,DC}\right|^{n-1} \cdot \sqrt{\sqrt[n]{2} - 1} \cdot GBW_{1\text{-}stage} . \end{aligned} \tag{4.81}$$

The evolution of $GBW_{cascade}$ normalized to $GBW_{1\text{-}stage}$ as a function of the number of stages is shown in Fig. 4.25 for three different values of $\left|A_{1\text{-}stage,DC}\right|$. It can be noticed that $GBW_{cascade}$ increases exponentially as a function of the number of stages if the DC voltage gain of every stage is significantly greater than 1. If $\left|A_{1\text{-}stage,DC}\right| = 1$, the normalized $GBW_{cascade}$ actually decreases when cascading more and more stages because the gain of the cascade does not increase while the bandwidth decreases. Therefore, cascading stages is only advantageous if the voltage gain of the composing stages is high enough. In contrast to what might be expected, there is no value for $\left|A_{1\text{-}stage,DC}\right|$ so that $GBW_{cascade} = GBW_{1\text{-}stage}$ for all values of n.

The noise of a cascade of amplifiers is a function of the number of stages, the topology and the sizing of each stage. Obviously, because more circuit elements are involved, the noise of the cascade is always higher than the noise of its individual stages. The output noise voltage spectral density of a cascade of n 1-stage amplifiers

---

[10] The value of C_i is the sum of the output capacitance of stage i and the input capacitance of stage $i + 1$. Therefore, the last stage is loaded with a capacitance that is as large as the input capacitance of every stage.

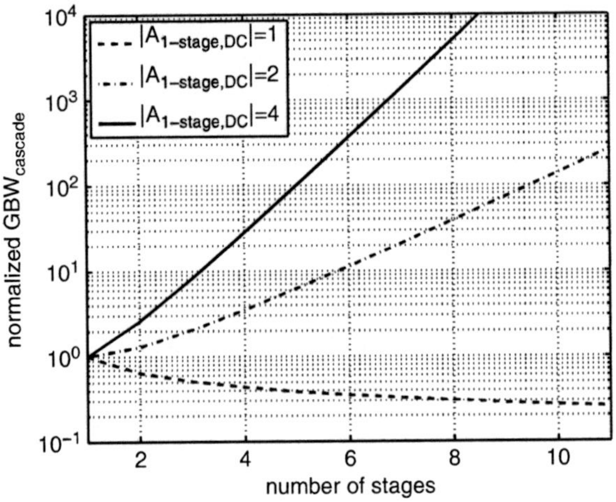

Fig. 4.25 Normalized $GBW_{cascade}$ as a function of the number of stages for three different values of $|A_{1\text{-}stage,DC}|$

is given by the following formula:

$$\overline{dv^2_{n,out,cascade}} = \overline{dv^2_{n,out,1}} \cdot |A_2 A_3 \ldots A_n|^2 + \overline{dv^2_{n,out,2}} \cdot |A_3 A_4 \ldots A_n|^2 + \cdots + \overline{dv^2_{n,out,n}}, \tag{4.82}$$

where $\overline{dv^2_{n,out,i}}$ and A_i represent the output noise voltage spectral density and the frequency dependent voltage gain of the ith stage, respectively. As expected, the noise of the first stages is amplified to a higher extent than the noise of the latter stages. This can also be seen when the output noise voltage spectral density is referred to the input:

$$\overline{dv^2_{n,in,cascade}} = \frac{\overline{dv^2_{n,out,1}}}{|A_1|^2} + \frac{\overline{dv^2_{n,out,2}}}{|A_1 A_2|^2} + \cdots + \frac{\overline{dv^2_{n,out,n}}}{|A_1 A_2 \ldots A_n|^2} = \overline{dv^2_{n,in,1}} + \frac{\overline{dv^2_{n,in,2}}}{|A_1|^2} + \cdots + \frac{\overline{dv^2_{n,in,n}}}{|A_1 A_2 \ldots A_{n-1}|^2}, \tag{4.83}$$

where $\overline{dv^2_{n,in,i}}$ represents the input-referred noise voltage spectral density of the ith stage. It can be concluded that the noise behavior of the cascade is almost completely determined by the noise of the first stage(s), if the voltage gain of the first stage(s) is high enough.

Until now, it has been assumed that all stages have an identical topology and sizing. However, from the point of view of both bandwidth and noise, it is advantageous to size all stages differently. By contrast with the upscaling mechanism which is generally used to drive a large capacitive load, downscaling can be used here to increase the bandwidth and to reduce the noise of the cascade. This is due to two reasons:

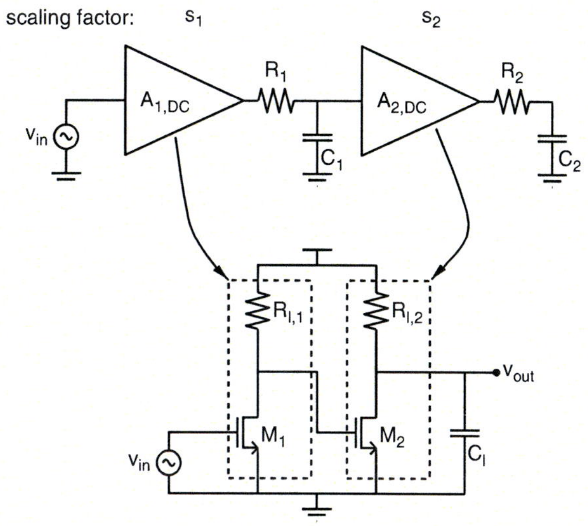

Fig. 4.26 A cascade of two common-source amplifiers with different scaling factors s_1 and s_2

1. Downscaling reduces the capacitive load at the output of the previous stage, which consequently results in a higher bandwidth;
2. Thanks to the downscaling mechanism, most of the current is steered through the first stages which mainly determine the noise of the cascade.

In order to examine this effect, consider the 2-stage amplifier in Fig. 4.26 in which both stages are implemented as common-source amplifiers, each with a different scaling factor. These scaling factors denote how much smaller or larger than 1 mS the transconductance of the respective input transistor is:

$$s_1 = \frac{g_{m,M_1}}{1\ \text{mS}}, \tag{4.84}$$

$$s_2 = \frac{g_{m,M_2}}{1\ \text{mS}}. \tag{4.85}$$

The output resistance of the input transistor decreases if the scaling factor is increased because the intrinsic gain of a transistor depends only on the technology used and the overdrive voltage, as discussed before, and not on the amount of current through it. Also, the resistance of the load resistor decreases when the scaling factor is raised because the DC voltage at the output, which is determined by the load resistor, needs to be maintained if the current is increased. The resistors R_1 and R_2 can be related with (4.58) to the circuit elements as follows:

$$R_1 = \frac{r_{ds,M_1} R_{l,1}}{r_{ds,M_1} + R_{l,1}} = \frac{1}{s_1} \cdot \frac{r_{ds} R_l}{r_{ds} + R_l}, \tag{4.86}$$

$$R_2 = \frac{r_{ds,M_2} R_{l,2}}{r_{ds,M_2} + R_{l,2}} = \frac{1}{s_2} \cdot \frac{r_{ds} R_l}{r_{ds} + R_l}, \tag{4.87}$$

where r_{ds} and R_l denote the output resistance of the input transistor and the load resistance, respectively, when the transconductance of the input transistor equals 1 mS. It can be derived from the formulas above that R_1 and R_2 decrease linearly with the scaling factor. As a result, the voltage gain of a single stage does not depent on the scaling factor. By contrast, the parasitic capacitances of a transistor increase linearly with the scaling factor because the physical sizes of the devices increase. With (4.65), the values of C_1 and C_2 can be expanded as follows:

$$\begin{aligned} C_1 &= C_{gd,M_1} + C_{gs,M_2} + \left(1 + \left|A_{cs,DC}\right|\right) C_{gd,M_2} \\ &= s_1 C_{gd} + s_2\left(C_{gs} + \left(1 + \left|A_{cs,DC}\right|\right) C_{gd}\right), \end{aligned} \tag{4.88}$$

$$\begin{aligned} C_2 &= C_{gd,M_2} + C_l \\ &= s_2 C_{gd} + C_l, \end{aligned} \tag{4.89}$$

where C_{gs} and C_{gd} denote the gate-source and gate-drain capacitance respectively of a transistor with a transconductance of 1 mS. Using (4.67) and (4.83), the input noise voltage spectral density of this 2-stage amplifier can be written as follows:

$$\begin{aligned} \overline{dv^2_{n,in,cascade}} &= \frac{1}{s_1} \cdot \left(\frac{4kTR_l}{|A_{cs}|^2} + \frac{4kT\gamma}{1\text{ mS}}\right) \cdot df \\ &+ \frac{1}{s_2} \cdot \frac{1}{|A_{cs}|^2} \cdot \left(\frac{4kTR_l}{|A_{cs}|^2} + \frac{4kT\gamma}{1\text{ mS}}\right) \cdot df. \end{aligned} \tag{4.90}$$

If the 2-stage amplifier with common-source stages represented in Fig. 4.26 is implemented in a 130 nm CMOS technology, the relationships in Fig. 4.20 can be used to calculate its performance. For an overdrive voltage of 200 mV, a supply voltage of 1.2 V and a DC output voltage of 0.6 V, the following values are applicable for a transconductance of 1 mS:

$$\begin{aligned} r_{ds} &= 11.6\text{ k}\Omega, \\ R_l &= 4.55\text{ k}\Omega, \\ C_{gs} &= 1.36\text{ fF}, \\ C_{gd} &= 0.413\text{ fF}, \end{aligned}$$

and the voltage gain of a single stage equals 3.27, as can be calculated with (4.58) and (4.59) independently of the value of the scaling factor. To illustrate the effect of downscaling on the performance of a 2-stage amplifier, two design choices are compared as an example:

1. Design 1: both stages have a scaling factor of 1;
2. Design 2: the first stage has a scaling factor of 1.5 and the second stage has a scaling factor of 0.5.

Table 4.2 Overview of the performance difference between a non-downscaled and a downscaled version of a 2-stage amplifier with common-source stages in a 130 nm CMOS process with a load capacitance of 1 fF

	Design 1	Design 2
$s_1 - s_2$	1 – 1	1.5 – 0.5
gain	10.7	10.7
bandwidth	12.7 GHz	16.5 GHz
input noise	5.64 $\text{nV}/\sqrt{\text{Hz}}$	4.99 $\text{nV}/\sqrt{\text{Hz}}$
current consumption	264 μA	264 μA

Both designs consume the same amount of current because the sum of the two scaling factors or, equivalently, the sum of both transconductances is the same in both cases. However, their performance is not the same, as can be concluded from Table 4.2. The gain for both designs is the same, namely $3.27^2 = 10.7$ because this does not depend on the scaling factor. For a fixed load capacitance of 1 fF for the two designs, it is no longer true that both poles coincide, which is required to calculate the bandwidth with (4.80). Although the second stage has a larger output resistance in the second design which needs to drive the same load capacitance, the overall bandwidth of the 2-stage amplifier is higher for the second design thanks to the pole on the intermediate node which is at a significantly higher frequency than in the first design. The input noise voltage spectral density of the first design equals 5.64 $\text{nV}/\sqrt{\text{Hz}}$, of which 5.40 $\text{nV}/\sqrt{\text{Hz}}$ is produced by the first stage and only 1.65 $\text{nV}/\sqrt{\text{Hz}}$ by the second stage. For the same current consumption, the second design has a significantly smaller input noise voltage spectral density of 4.99 $\text{nV}/\sqrt{\text{Hz}}$, of which 4.41 $\text{nV}/\sqrt{\text{Hz}}$ originates from the first stage and 2.33 $\text{nV}/\sqrt{\text{Hz}}$ from the second stage.

4.3.2 Design of the Feedback Network

The feedback network of the shunt-shunt feedback TIA has until now been assumed to be composed of a single resistor, as can be seen in Fig. 4.14. While several possibilities exist to implement this resistor in a CMOS process, the most straightforward manner is to use a polysilicon resistor. This is actually a piece of polycrystalline silicon with a predefined resistance per square. The resistance can be set accurately by choosing an appropriate length and width. A large resistance is obtained if the width is minimized, typically around 1 μm in a contemporary CMOS process, and the length is maximized, typically several tens of micrometers for resistances between 10 and 100 kΩ. It is shown in [Kno03] that such a structure has strong capacitive coupling to the substrate. Consequently, this kind of resistor contributes significantly to the frequency response of the TIA of which it is part. According to [Kno03], this may result in bandwidth limitations, signal distortion, and instability.

The equivalent circuit of a polysilicon resistor, as proposed in [Kno03], is shown in Fig. 4.27a. The wanted resistance is represented by R_1, whereas the other components are added to model the non-ideal behavior of such a resistor at high frequencies. Their

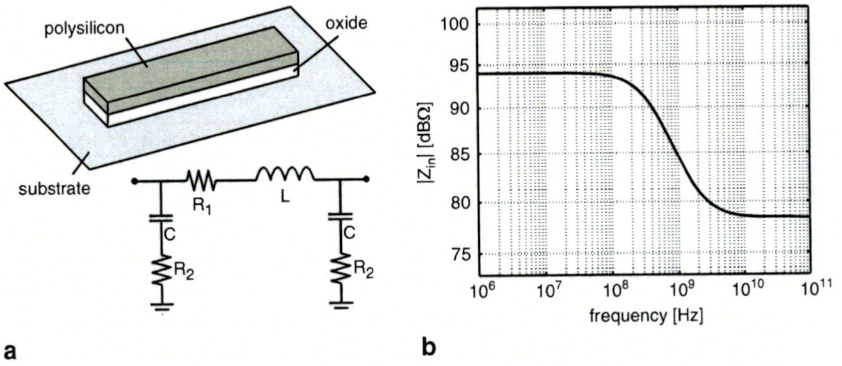

Fig. 4.27 The non-idealities of a polysilicon resistor. **a** Layout and equivalent circuit of a polysilicon resistor [Kno03]. **b** Magnitude of the input impedance of a polysilicon resistor when the other side is grounded

values are given by the following formulas [Kno03]:

$$R_1 = R' \cdot l, \tag{4.91}$$

$$R_2 = \frac{R' \cdot l}{6}, \tag{4.92}$$

$$C = \frac{C' \cdot l}{2}, \tag{4.93}$$

$$L = \frac{R'^2 \cdot C' \cdot l^3}{6}, \tag{4.94}$$

where R' and C' represent the resistance and capacitance per unit length and l is the physical length of the resistor. The values of R' and C' depend on the width of the polysilicon resistor. While the current flows completely through R_1 and L at low frequencies, experiencing an impedance approximately equal to R_1, at very high frequencies the current flows through the nearest capacitor and the connected resistor with a value R_2. Typical values for R' and C' are 1 kΩ/μm and 0.2 fF/μm, respectively, if the width is 1 μm [Kno03]. With the formulas above, the following results are obtained for a length of 50 μm: $R_1 = 50$ kΩ, $R_2 = 8.33$ kΩ, $C = 5.00$ fF and $L = 4.17$ μH. The effect of these considerably large parasitic resistance, capacitance and inductance can be visualized by observing the input impedance that is seen when looking into one of the terminals of this polysilicon resistor while the other terminal is grounded. As can be seen in Fig. 4.27b, the input impedance is determined by the value of R_1 at low frequencies, which is equal to 50 kΩ = 94.0 dBΩ in this example. At very high frequencies, the input impedance is completely determined by the value of R_2, which is equal to 8.33 kΩ = 78.4 dBΩ in this case. Capacitive behavior can be observed for frequencies between approximately 100 MHz and 10 GHz. Therefore, such a polysilicon resistor cannot be used as the feedback resistor of a shunt-shunt feedback TIA if the desired bandwidth is in the GHz-range.

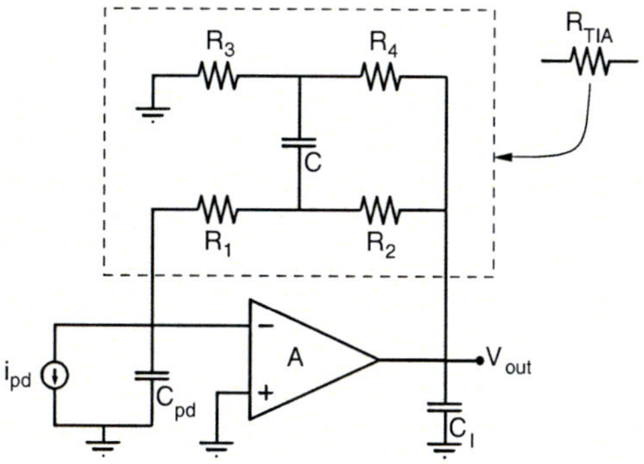

Fig. 4.28 A shunt-shunt feedback TIA with a compound feedback network to alleviate the problems that arise when integrating a polysilicon resistor with a high value [Sei03]

A first solution to this problem is shown in Fig. 4.28, as proposed in [Sei03] and applied in [Sei04]. The polysilicon feedback resistor R_{TIA} of a shunt-shunt feedback TIA can be replaced by the network of four resistors and one capacitor. The values of the resistors should obey the following relationships:

$$R_1 + R_2 = R_{TIA}, \tag{4.95}$$

$$R_2 = \alpha R_1, \tag{4.96}$$

$$R_3 = \frac{R_1}{\beta}, \tag{4.97}$$

$$R_4 = \frac{R_2}{\beta}, \tag{4.98}$$

where α and β can be chosen freely. The concept of the feedback network is its different behavior at low and high frequencies. At low frequencies, the transimpedance of the TIA in Fig. 4.28 is simply the sum of R_1 and R_2. If the open-loop voltage gain A is sufficiently high, the input node is a virtual ground[11]. As a result, the voltage drops across R_1 and R_2 are the same as the voltage drops across R_3 and R_4 since both voltage dividers are related as shown in (4.97) and (4.98). Consequently, the voltages at both terminals of C are equal to each other and, taking into account the relationship (4.96), they are equal to $1/(1+\alpha)$ times the voltage at the output node. At high frequencies, the capacitor C behaves like as a short circuit and R_4 appears in parallel with R_2. If a large value of β is chosen, the resistor R_2 has no effect at these high frequencies as it is in parallel with a resistor R_4 that is β times smaller. Consequently, the parasitics of R_2 have no effect on the behavior of the TIA. Only the parasitics of R_1 are of importance but, if α is made large enough, these parasitics are significantly smaller than those of R_2 or R_{TIA}. The transimpedance at these high frequencies is the same as at low frequencies because the same voltage drop is realized across R_1, which means that the same current is fed back to the input.

[11] A virtual ground is a node of a circuit that is maintained at a steady voltage without being connected directly to that voltage.

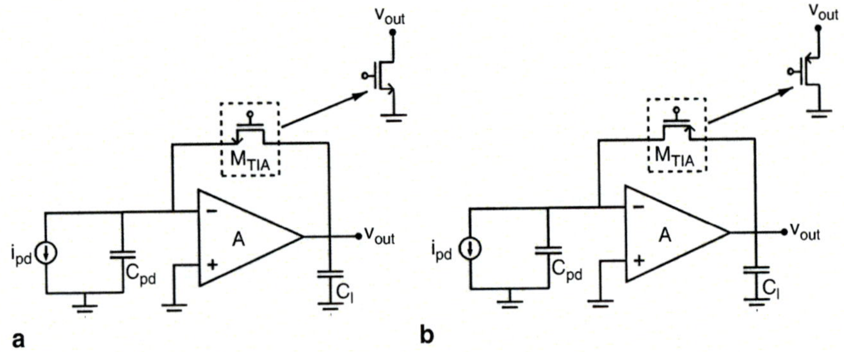

Fig. 4.29 A shunt-shunt feedback TIA with a transistor in its linear region as feedback device [Ing04]. **a** An nMOS has its source at the input node. **b** An nMOS has its source at the output node

Another solution to circumvent the problems that arise when integrating a polysilicon resistor with a large resistance is to replace it by a transistor in its linear region, as shown in Fig. 4.29 [Ing04]. In this region ($V_{ds} < V_{gs} - V_{th}$), the transistor acts as a resistor with the following value:

$$R_{M_{TIA}} = \begin{cases} \dfrac{1}{\mu_{e^-} C_{ox} \frac{W}{L} \left(V_{gs} - V_{th}\right)} & \text{for an nMOS} \\ \left| \dfrac{1}{\mu_{h^+} C_{ox} \frac{W}{L} \left(V_{gs} - V_{th}\right)} \right| & \text{for a pMOS,} \end{cases} \tag{4.99}$$

where C_{ox} is the gate-oxide capacitance per unit area and W and L are the width and length of transistor M_{TIA}, respectively. From the formulas above, it can be concluded that a large resistance is realized for a low overdrive voltage, a short width W, a long length L or a combination of those. As a result, the physical size of the feedback transistor can be kept very small, especially if it should show a large equivalent resistance. Consequently, the parasitic capacitances are negligible, which makes this an ideal solution when a high bandwidth is required.

It is possible to use an nMOS as well as a pMOS in the feedback path, as shown in Fig. 4.29a, b, respectively. If an nMOS is used, the source is at the input node which is a virtual ground because the photodiode current flows out of the TIA, from the cathode to the anode of the photodiode. As a consequence, the overdrive voltage remains constant while the drain-source voltage increases with an increasing current. The feedback transistor can thus enter the saturation region for large input currents where the resistance is significantly higher. This situation should be avoided. By contrast, if a pMOS is used, the source is at the output node while the drain is at the input node. If a current flows from the photodiode cathode to its anode, the output voltage rises and the overdrive voltage rises accordingly. Any change in V_{ds} is reflected in an equal change in V_{gs} so that M_{TIA} can never enter the saturation region. A pMOS is therefore preferred.

4.3.3 *The Capacitance-Relieved TIA*

As shown in the previous chapter, the parasitic capacitance of an integrated photodiode is very large. Consequently, several circuit techniques have been proposed in the literature to alleviate the problem this poses when designing a high-speed optical receiver. One of these is the so-called *self-compensated TIA*[12] proposed in [Tsa04, Tsa05, Tsa09] and shown in Fig. 4.30a. This technique is used as a reference against which to compare the newly developed *capacitance-relieved TIA*, which is shown in Fig. 4.30b. To simplify the analysis, it is assumed that no load capacitance is present at the output node of either topology.

The self-compensated TIA consists of two feedback loops. The first loop comprises resistor R_{TIA}, which is used as the negative feedback path over an amplifier with voltage gain A. Due to the inverting amplification across the resistor, the input

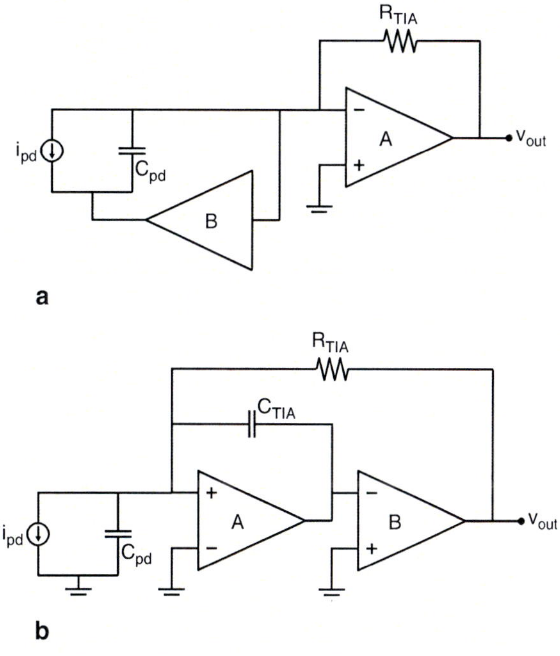

Fig. 4.30 Circuit solutions to increase the transimpedance-bandwidth product of a shunt-shunt feedback TIA. **a** The self-compensated TIA [Tsa04, Tsa05, Tsa09]. **b** The capacitance-relieved TIA

[12] In the literature, this technique is generally known as *bootstrapping* which has been introduced to nullify the effect of the input capacitance of a coaxial cable [San06].

resistance is a factor A lower than R_{TIA} thereby increasing the pole at the input node by the same amount, like in a regular shunt-shunt feedback TIA. The second feedback loop exists around the amplifier with gain B. Its function is to make sure that the transient voltage across the photodiode is reduced. This reduced voltage swing results in a smaller effective input capacitance. The resistance and capacitance that appear at the input node and determine the bandwidth of the TIA are thus both reduced by the first and second loop, respectively, and the bandwidth of the TIA is consequently increased. The transfer function of the self-compensated TIA is given by the following formula:

$$\begin{aligned} Z_{SCTIA} &= \frac{AR_{TIA} - R_{out,A}}{1 + A + j2\pi f C_{pd}((1 + A)R_{out,B} + (1 - B)(R_{out,A} + R_{TIA}))} \\ &\approx \frac{R_{TIA}}{1 + j2\pi f C_{pd}\left(R_{out,B} + R_{TIA}\left(\frac{1-B}{A}\right)\right)}, \end{aligned} \tag{4.100}$$

where it is assumed in the approximation that A is much larger than 1 and that $R_{out,A}$ is much smaller than R_{TIA}. The effect of the second feedback loop is clearly visible in (4.100). If B equals 1, which is supposed from now on, the bandwidth of the self-compensated TIA is determined only by $R_{out,B}$ and C_{pd}. The DC transimpedance gain and bandwidth are then given by:

$$|Z_{SCTIA}| = R_{TIA}, \tag{4.101}$$

$$BW_{SCTIA} = \frac{1}{2\pi R_{out,B} C_{pd}}. \tag{4.102}$$

Consequently, a unity gain amplifier with a low output resistance $R_{out,B}$ is needed to achieve a high bandwidth, while a high transimpedance gain is achieved by making R_{TIA} large. By contrast with the normal shunt-shunt feedback TIA, this has no effect on the bandwidth.

The self-compensated TIA technique cannot be adopted when the photodiode is integrated together with the circuit in a CMOS process because the anode is then always grounded. The voltage at the anode can therefore never be altered, which is a necessity for the self-compensated TIA to operate. The capacitance-relieved TIA, shown in Fig. 4.30b, solves this problem. Again two feedback loops can be distinguished. The first loop is composed of two amplifiers and resistor R_{TIA}. To make the voltage gain across R_{TIA} inverting, one of the amplifiers is inverting while the other one is non-inverting. The objective of the inverting amplification is again to reduce the input resistance by a factor AB. The second feedback loop is nested inside the first one and consists of capacitor C_{TIA} and a non-inverting amplifier with gain A. The function of this loop is to provide a negative capacitance at the input of the TIA. This negative capacitance appears in parallel with the photodiode capacitance. The result is that the effective input capacitance is decreased so that the bandwidth of the TIA is increased accordingly. The transfer function of the capacitance-relieved TIA is given by:

$$Z_{CRTIA} \approx \frac{R_{TIA}\left(1 + j2\pi f \frac{C_{TIA}}{A} R_{out,A}\right)}{1 + j2\pi f R_{TIA}\left(\frac{C_{pd}}{AB} - \frac{C_{TIA}}{B}\right) - 4\pi^2 f^2 \frac{R_{TIA} R_{out,A} C_{pd} C_{TIA}}{AB}}, \tag{4.103}$$

where it is assumed that $R_{out,A}$ and $R_{out,B}$ are negligible compared to R_{TIA} and that A is much greater than 1. When B equals 1 and if it is assumed that the two poles in (4.103) are far from each other, the DC transimpedance and the bandwidth of this TIA are given by:

$$|Z_{CRTIA}| = R_{TIA}, \tag{4.104}$$

$$BW_{CRTIA} = \frac{1}{2\pi R_{TIA}\left(\frac{C_{pd}}{A} - C_{TIA}\right)}, \tag{4.105}$$

respectively. It can be seen in the formula above that BW_{CRTIA} becomes infinite by choosing C_{TIA} a factor A smaller than C_{pd}. However, in this case the two poles are not far from each other and the bandwidth is no longer defined by (4.105). The zero in (4.103) is at a very high frequency if $R_{out,A}$ is small and A is large. It is therefore neglected from here on. The DC transimpedance of the capacitance-relieved TIA is identical to that of the self-compensated TIA if R_{TIA} has the same value in both topologies.

Three noise contributors can be distinguished in the self-compensated TIA in Fig. 4.30a: R_{TIA}, and the two voltage amplifiers. If, as before, only white noise is taken into account, the output noise voltage spectral density of the self-compensated TIA is given by:

$$\overline{dv^2_{n,out,SCTIA}} = \overline{dv^2_{n,R_{TIA}}} + \overline{dv^2_{n,A}} + \overline{dv^2_{n,B}} \cdot \left|\frac{j2\pi f R_{TIA} C_{pd}}{1 + \frac{jf}{BW_{SCTIA}}}\right|^2, \tag{4.106}$$

for frequencies below BW_{SCTIA}, where it is again assumed that $R_{out,A}$ is small compared to R_{TIA}, A is much larger than 1 and $B = 1$. In this formula, $\overline{dv^2_{n,A}}$ and $\overline{dv^2_{n,B}}$ represent the input-referred noise voltage spectral densities of the two voltage amplifiers. The three noise components in (4.106) are depicted in Fig. 4.31a, where it is assumed that $\overline{dv^2_{n,B}} > \overline{dv^2_{n,A}}$, which is a reasonable assumption because a unity gain buffer does not suppress the noise of the devices after the input transistor. To obtain the total output noise voltage, (4.106) should be integrated over the entire frequency spectrum. However, as the output capacitance, and thus the pole at the output, has been neglected to simplify the calculations, this would lead to an infinite RMS output noise voltage, which is clearly not correct. To obtain a good approximation of the RMS output noise voltage, (4.106) is integrated from 0 Hz up to the bandwidth of

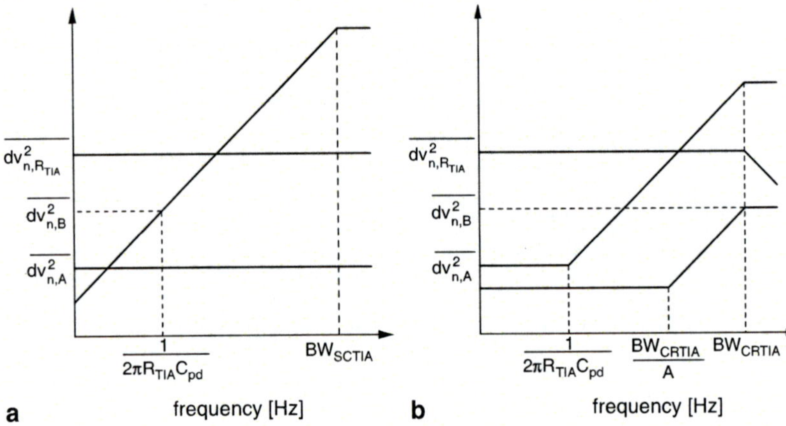

Fig. 4.31 Noise voltage spectral density at the output of the self-compensated TIA and the capacitance-relieved TIA. **a** The self-compensated TIA. **b** The capacitance-relieved TIA

the self-compensated TIA, which results in the following formula:

$$v_{n,out,RMS,SCTIA} = \sqrt{\int_0^{BW_{SCTIA}} \overline{dv^2_{n,out,SCTIA}}} \approx \sqrt{4kTBW_{SCTIA}\left(R_{TIA} + \frac{\gamma}{g_{m,A}} + \frac{\gamma}{g_{m,B}}\frac{R^2_{TIA}}{4R^2_{out,B}}\right)}, \quad (4.107)$$

where $g_{m,A}$ and $g_{m,B}$ are the transconductances that generate the same amount of noise as both voltage amplifiers. Because A is significantly larger than 1, the value of $g_{m,A}$ is approximately equal to the transconductance of the input transistor of this voltage amplifier. By contrast, the value of $g_{m,B}$ is significantly larger than the transconductance of its input transistor because no gain is available to suppress the noise of the other devices, as already mentioned. It can be concluded from (4.107) and Fig. 4.31a that the unity gain amplifier is the primary contributor to the total noise voltage at the output. The RMS output noise voltage can also be referred to the input as follows:

$$i_{n,in,RMS,SCTIA} = \frac{v_{n,out,RMS,SCTIA}}{R_{TIA}} = \sqrt{4kTBW_{SCTIA}\left(\frac{1}{R_{TIA}} + \frac{\gamma}{g_{m,A}R^2_{TIA}} + \frac{\gamma}{4g_{m,B}R^2_{out,B}}\right)}. \quad (4.108)$$

The same three noise contributors can be distinguished for the capacitance-relieved TIA in Fig. 4.30b. For frequencies up to the bandwidth BW_{CRTIA}[13], the output noise

[13] Poles and zeros above the bandwidth of the capacitance-relieved TIA are neglected.

voltage spectral density of this topology is given by:

$$\begin{aligned}\overline{dv^2_{n,out,CRTIA}} &= \overline{dv^2_{n,R_{TIA}}} \cdot \left|\frac{1}{1+\frac{jf}{BW_{CRTIA}}}\right|^2 \\ &+ \overline{dv^2_{n,A}} \cdot \left|\frac{1+j2\pi f R_{TIA} C_{pd}}{1+\frac{jf}{BW_{CRTIA}}}\right|^2 \\ &+ \frac{\overline{dv^2_{n,B}}}{A^2} \cdot \left|\frac{1+\frac{jfA}{BW_{CRTIA}}}{1+\frac{jf}{BW_{CRTIA}}}\right|^2 .\end{aligned} \tag{4.109}$$

The three noise components in this formula are shown in Fig. 4.31b where, to make a fair comparison, it is assumed that the value of R_{TIA} and the designs of both voltage amplifiers are identical to those that were used in the self-compensated TIA of Fig. 4.31a. It can be noticed immediately that the capacitance-relieved TIA has a lower output noise voltage spectral density which will be confirmed later. Integrating (4.109) between 0 Hz and BW_{CRTIA} results in the RMS output noise voltage of this topology:

$$\begin{aligned}v_{n,out,CRTIA} &= \sqrt{\int_0^{BW_{CRTIA}} \overline{dv^2_{n,out,CRTIA}}} \\ &\approx \sqrt{4kTBW_{CRTIA}} \\ &\times \sqrt{\left(R_{TIA}\frac{\pi}{4} + \frac{\gamma}{g_{m,A}}\pi^2 R^2_{TIA} C^2_{pd} BW^2_{CRTIA} + \frac{\gamma}{g_{m,B}}\frac{A^2}{4}\right)},\end{aligned} \tag{4.110}$$

where $g_{m,A}$ and $g_{m,B}$ again represent the equivalent transconductances of the voltage amplifiers. By contrast with the self-compensated TIA, in which the unity gain amplifier dominates the output noise voltage, the RMS output noise voltage of the capacitance-relieved TIA is determined mainly by the noise of the voltage amplifier with amplification A, as can be concluded from (4.110) and Fig. 4.31b. The RMS input noise current of this topology is as follows:

$$\begin{aligned}i_{n,in,RMS,CRTIA} &= \frac{v_{n,out,RMS,CRTIA}}{R_{TIA}} \\ &= \sqrt{4kTBW_{CRTIA}\left(\frac{\pi}{4R_{TIA}} + \frac{\gamma}{g_{m,A}}\pi^2 C^2_{pd} BW^2_{CRTIA} + \frac{\gamma A^2}{4g_{m,B}R^2_{TIA}}\right)}.\end{aligned} \tag{4.111}$$

As in every feedback system, stability should be verified for the capacitance-relieved TIA[14]. Only a single loop gain needs to be considered because the two feedback

[14] The stability of the self-compensated TIA depends on two feedback loops which are not nested. Although this issue is not discussed in detail, it has to be stressed that B should never be greater than 1.

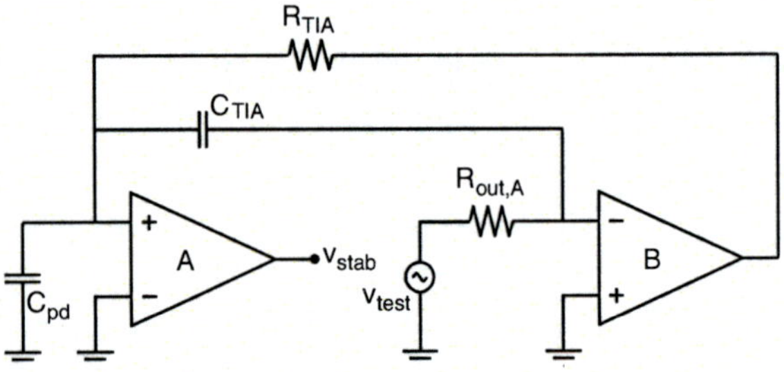

Fig. 4.32 Circuit diagram to analyze the stability of the capacitance-relieved TIA

loops of the topology are nested and can therefore be cut as shown in Fig. 4.32. The loop gain LG_{CRTIA} is as follows:

$$LG_{CRTIA} = \frac{v_{stab}}{v_{test}} \approx -A \cdot \frac{1 - j2\pi f R_{TIA} C_{TIA}}{1 + j2\pi f R_{TIA} C_{pd} - 4\pi^2 f^2 R_{TIA} R_{out,A} C_{pd} C_{TIA}}, \quad (4.112)$$

where it is again assumed that $R_{out,A}$ and $R_{out,B}$ **are significantly smaller than** R_{TIA} and that $B = 1$. To make the approximation in (4.112), it is also assumed that C_{TIA} has **a significantly lower value than** C_{pd}**, which will soon be confirmed as a necessary** condition to guarantee a stable operation. It can be observed in (4.112) that the loop gain of the capacitance-relieved TIA contains two poles and a positive zero so that stability can indeed become a problem. The second pole, which is determined by $R_{out,A}$ and C_{TIA}, can be ignored because it is located at a very high frequency. However, this is not the case for the positive zero. The Bode plot of the loop gain in (4.112) is shown schematically in Fig. 4.33. Although the second pole can be ignored, this topology can become unstable because a positive zero shifts the phase by −90° just as a pole would do, while the magnitude is pushed upwards at 20 dB/decade just like a negative zero. The possibility therefore exists that the magnitude of the

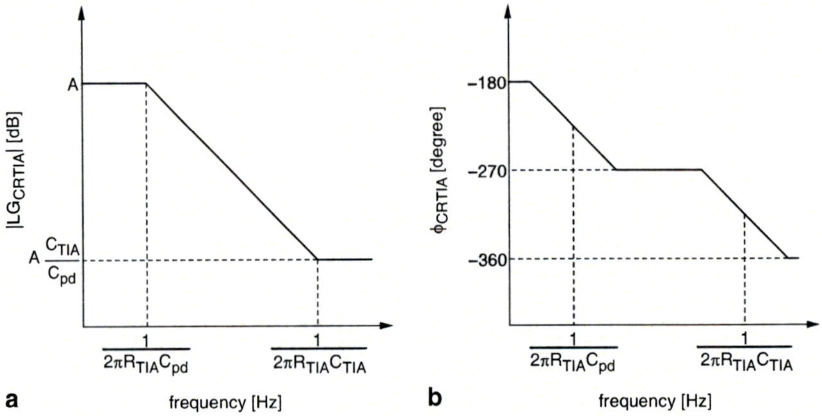

Fig. 4.33 Bode plot of the loop gain of the capacitance-relieved TIA. **a** Magnitude. **b** Phase

loop gain is larger than 1 at the frequency where the phase has shifted by $-180°$ so that the system becomes unstable. In order to prevent this, the following condition needs to be fulfilled:

$$A \cdot \frac{C_{TIA}}{C_{pd}} < 1 \Longleftrightarrow C_{TIA} < \frac{C_{pd}}{A}, \tag{4.113}$$

which means that the magnitude of the loop gain is below 1 for frequencies above the positive zero. Otherwise stated, this condition ensures that the effective capacitance at the input node of the capacitance-relieved TIA is positive.

Suppose, for example, that a self-compensated TIA and a capacitance-relieved TIA need to have a bandwidth of 1 GHz for a photodiode capacitance of 1 pF. If the desired DC transimpedance is 100 kΩ, this is also the resistance of both feedback resistors (4.101)–(4.104). The value of $R_{out,B}$ needs to be 159 Ω in order to obtain the bandwidth of 1 GHz for the self-compensated TIA (4.102). The same output resistance is assumed for the other voltage amplifier having a gain A of 100. The equivalent transconductance of each voltage amplifier depends on the power consumption and the internal architecture. The following values are used here: $g_{m,A} = 10$ mS and $g_{m,B} = 5$ mS. In order to achieve the same bandwidth as the self-compensated TIA with the same two voltage amplifiers and the same feedback resistor, C_{TIA} should equal 8.41 fF, as can be calculated with (4.105). The RMS input noise current of the resulting self-compensated TIA can be revealed with (4.108), namely 208 nA_{RMS} of which 12.9 nA_{RMS} is due to R_{TIA}, 470 pA_{RMS} comes from $g_{m,A}$ and 208 nA_{RMS} is contributed by $g_{m,B}$. By contrast, the capacitance-relieved TIA obtains an RMS input noise current of only 152 nA_{RMS} (4.111) of which 11.4 nA_{RMS} is caused by R_{TIA}, 148 nA_{RMS} comes from $g_{m,A}$ and 33.2 nA_{RMS} is contributed by $g_{m,B}$.

To conclude, it can be stated that, by contrast with the self-compensated TIA, the capacitance-relieved TIA is compatible with integrated photodiodes in CMOS and it also achieves a lower RMS input noise current for the same power consumption.

4.4 TIA Designs

To finalize this chapter, some TIA designs are discussed below. Almost all of them have also been realized as part of an integrated optical receiver. Only the design choices and the simulation results are treated here. The measurement results are presented in Chap. 7.

4.4.1 A Differential Shunt-Shunt Feedback TIA

As discussed in the previous chapter, a differential n-well/p-sub photodiode is able to provide a significantly higher bandwidth than a regular n-well/p-sub photodiode. Such a photodiode generates two currents, originating from the illuminated and the non-illuminated junctions respectively, which need to be subtracted from each other in order to obtain a fast response. Two methods can be applied to take care of this task:

1. Two single-ended TIAs convert the currents of the differential photodiode into voltages which are subtracted from each other by a dedicated difference amplifier;
2. One differential TIA converts the differential-mode component of the photocurrents into a differential voltage.

As mentioned in [Her07], it is very important to ensure that sufficient common-mode rejection is experienced by the photodiode currents since they contain a large common-mode component and a relatively small differential-mode component. It is therefore advantageous to reject the common-mode component throughout the complete receiver chain and consequently to implement as many of the building blocks as possible differentially. The second design methodology is therefore applied in this work.

The architecture of the differential shunt-shunt feedback TIA is depicted in Fig. 4.34. Four stages are cascaded to realize high voltage gain and high bandwidth. Each stage is implemented as a differential version of the common-source amplifier, as shown in Fig. 4.34. The feedback paths are realized by two identical pMOS transistors operating in their linear region.

In Fig. 4.34, the differential photodiode is represented by two current sources and two capacitances: $i_{pd,i}$ and $C_{pd,i}$ model the behavior of the illuminated photodiode,

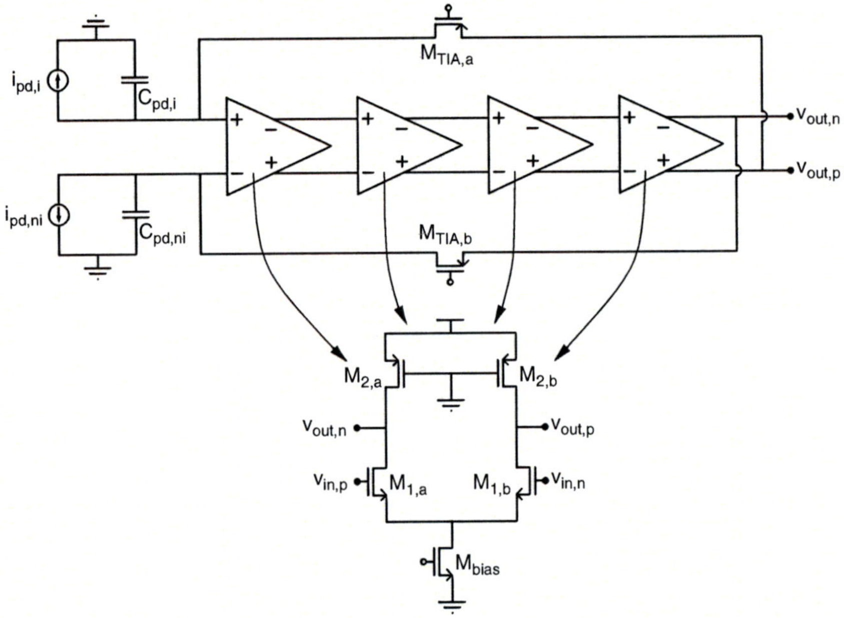

Fig. 4.34 Architecture of the differential shunt-shunt feedback TIA in 130 nm CMOS with four differential common-source amplifiers and linear pMOS transistors as the feedback network

whereas $i_{pd,ni}$ and $C_{pd,ni}$ do the same for the non-illuminated photodiode. While the two capacitances have the same value, namely 1.02 pF for a differential n-well/p-sub photodiode with a diameter of 60 μm in the 130 nm CMOS technology used, this is not the case for both photodiode currents. As discussed in the previous chapter, the current of the illuminated photodiode has a fast drift component and a slow diffusion component. By contrast, the current of the non-illuminated photodiode has only a slow diffusion component with the same magnitude as the slow component of the illuminated photodiode. The photodiode current can consequently be decomposed into differential-mode and common-mode components, as follows:

$$i_{pd,DM} = \underbrace{i_{pd,fast} + \frac{i_{pd,slow}}{2}}_{i_{pd,i}} - \underbrace{\frac{i_{pd,slow}}{2}}_{i_{pd,ni}} = i_{pd,fast} \tag{4.114}$$

$$i_{pd,CM} = \frac{i_{pd,i} + i_{pd,ni}}{2} = \frac{i_{pd,fast} + i_{pd,slow}}{2}, \tag{4.115}$$

where $i_{pd,fast}$ and $i_{pd,slow}$ are the fast and slow components, respectively. It can be concluded from the formulas above that the differential TIA should amplify the differential photocurrent, $i_{pd,DM}$, while suppressing the common-mode component, $i_{pd,CM}$.

The differential common-source amplifier has pMOS transistors operating in their linear region as loads, as can be seen in Fig. 4.34. The gates of these transistors are tied to ground in order to minimize their physical size for a given small-signal resistance. Apart from that single fact, this circuit has the same performance as the single-ended common-source amplifier (Sect. 4.3.1) only for differential-mode input signals. Common-mode input signals are suppressed thanks to the current source M_{bias}. In reality, however, a fraction of the common-mode component is transferred to the output due to the non-idealities of M_{bias}, namely its finite output resistance and the capacitance between its drain and ground. The differential-mode and common-mode DC voltage gains of the differential common-source amplifier are given respectively by:

$$\left|A_{DM,DC}\right| = g_{m,M_1} R_{out,DM}, \tag{4.116}$$

$$\left|A_{CM,DC}\right| \approx \frac{R_{M_2}}{2r_{ds,M_{bias}}}, \tag{4.117}$$

where g_{m,M_1} denotes the transconductance of the input transistors, $R_{out,DM}$ is the small-signal output resistance for differential-mode output signals, R_{M_2} is the equivalent resistance of the linear pMOS loads, as defined by (4.99), and $r_{ds,M_{bias}}$ is the small-signal output resistance of M_{bias}. The output resistance of the differential common-source amplifier is different for differential-mode and common-mode signals, namely:

$$R_{out,DM} = \frac{r_{ds,M_1} R_{M_2}}{r_{ds,M_1} + R_{M_2}}, \tag{4.118}$$

$$R_{out,CM} \approx R_{M_2}. \tag{4.119}$$

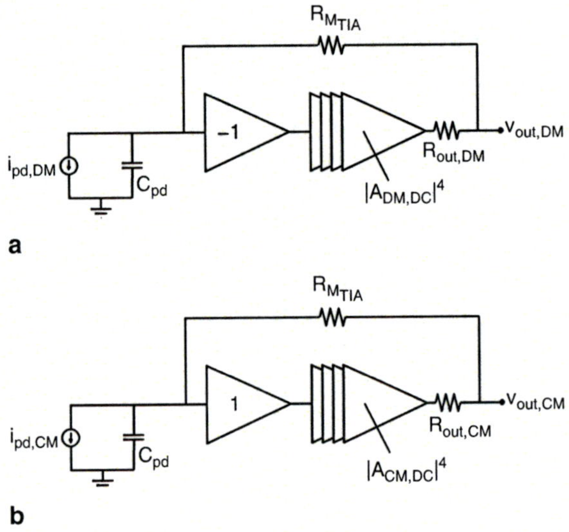

Fig. 4.35 Model of the differential shunt-shunt feedback TIA. **a** Model for differential-mode input currents. **b** Model for common-mode input currents

These formulas show that the output resistance of the differential common-source amplifier is larger for common-mode signals than for differential-mode signals.

To analyze the behavior of the differential shunt-shunt feedback TIA shown in Fig. 4.34 for differential-mode and common-mode input currents, the models of Fig. 4.35 are applied. As can be seen in Fig. 4.35a, an inversion is needed to account for the fact that the feedback is negative. For common-mode signals, this is not needed because the common-mode output voltage is always in-phase with the common-mode input voltage in the applied 4-stage amplifier. If the output capacitance and the finite bandwidth of the voltage amplifier are neglected, the transimpedance of the differential shunt-shunt feedback TIA for differential-mode input currents is given by:

$$Z_{TIA,DM} = \frac{\left|A_{DM,DC}\right|^4 R_{M_{TIA}} - R_{out,DM}}{1 + \left|A_{DM,DC}\right|^4 + j2\pi f\left(R_{M_{TIA}} + R_{out,DM}\right) C_{pd}} \approx \frac{R_{M_{TIA}}}{1 + j2\pi f \frac{R_{M_{TIA}} C_{pd}}{\left|A_{DM,DC}\right|^4}}, \quad (4.120)$$

where it is assumed that $R_{out,DM}$ is significantly smaller than $R_{M_{TIA}}$, the equivalent resistance of M_{TIA}, and that $\left|A_{DM,DC}\right|^4$ is much larger than 1. For differential-mode input currents, the transimpedance of this topology is similar to that of the single-ended shunt-shunt feedback TIA, as discussed in Sect. 4.2. Common-mode input currents are subjected to a different transimpedance gain, as can be seen in the following formula:

$$Z_{TIA,CM} = \frac{\left|A_{CM,DC}\right|^4 R_{M_{TIA}} + R_{out,CM}}{1 - \left|A_{CM,DC}\right|^4 + j2\pi f \left(R_{M_{TIA}} + R_{out,CM}\right) C_{pd}} \approx \frac{R_{out,CM}}{1 + j2\pi f R_{M_{TIA}} C_{pd}}, \tag{4.121}$$

where it is assumed that $R_{out,CM}$ is significantly smaller than $R_{M_{TIA}}$ and that $\left|A_{CM,DC}\right|^4$ is much smaller than 1. Note that the DC transimpedance gain for common-mode input currents is determined only by the common-mode output resistance of the last amplifier stage. The bandwidth for common-mode input currents is $\left|A_{DM,DC}\right|^4$ times smaller than for differential-mode input currents, as can be concluded by comparing (4.120) and (4.121).

The noise of a shunt-shunt feedback TIA is determined primarily by the value of the feedback resistor and the transconductance of the first stage of the voltage amplifier, as discussed in Sect. 4.2. To obtain an output voltage with a sufficiently large swing, the equivalent resistance of the linear pMOS transistors is chosen equal to 5 kΩ in this design. The stages of the voltage amplifier are subjected to the principle of downscaling, a technique that has been described in Sect. 4.3.1. The transconductance of the input transistors of the first amplifier stage equals 229 mS, which is optimal in terms of noise. For a lower transconductance, the noise of the complete TIA rises, reducing the SNR. For a higher value, the input capacitance increases; this forces the value of the feedback resistance to decrease in order to maintain the same bandwidth. This again reduces the SNR. The transconductance of the other three stages is 10 times smaller, namely 22.9 mS. This is done primarily to balance the noise of the first stage with that of the subsequent stages and to save power as a result. The overdrive voltage of the input transistors is 200 mV for all stages. As the design is made in a 130 nm CMOS process, the properties of Fig. 4.20 can be applied here once again. The properties of the input transistors of the four amplifier stages for this first design are summarized in Table 4.3.

As can be concluded from Table 4.3, the total current consumption in this first design is dominated by the current consumption of the first amplifier stage which is needed to minimize the circuit noise. However, the white noise of a transistor is determined by its transconductance and not by its drain-source current. In the second design, the overdrive voltage of the input transistor of the first stage is only 50 mV

Table 4.3 Comparison of the equivalent single-ended properties of the input transistors of the designed voltage amplifiers

	Design 1		Design 2	
	Stage 1	Stage 2–4	Stage 1	Stage 2–4
$V_{gs} - V_{th}$	200 mV	200 mV	50 mV	200 mV
g_m	229 mS	22.9 mS	113 mS	22.9 mS
C_{gs}	320 fF	32.0 fF	320 fF	32.0 fF
I_{ds}	29.8 mA	2.98 mA	6.81 mA	2.98 mA
I_{tot}	38.7 mA		15.8 mA	

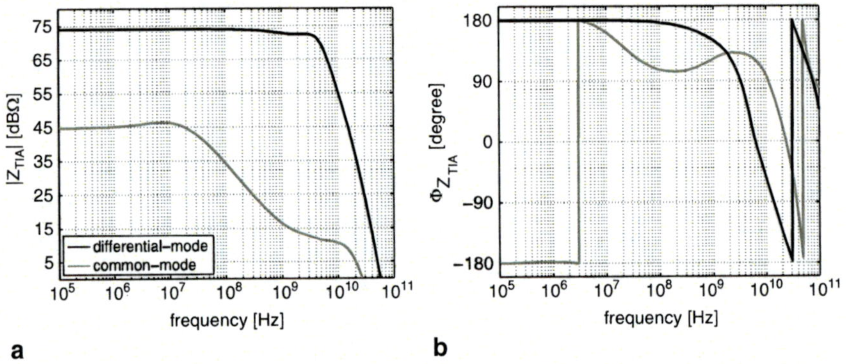

Fig. 4.36 Simulated Bode plot of the differential shunt-shunt feedback TIA for differential-mode and common-mode input currents. **a** Magnitude. **b** Phase

instead of 200 mV in the first design. The three other stages are identical as in the first design. This leads to the values that are listed in Table 4.3. Note that for the same C_{gs} as in the first design, the transconductance of the input transistor of the first stage in the second design is a factor of 2 lower while the drain-source current is more than 4 times lower.

As mentioned already, the equivalent resistance of the linear pMOS feedback transistors is chosen equal to 5 kΩ, which results in a DC transimpedance for differential-mode input currents of 73.9 dBΩ = 4.95 kΩ. This is confirmed by the Bode plot in Fig. 4.36. A differential photodiode with a parasitic capacitance of 1.02 pF is connected at the input[15]. If the first design which has a DC voltage gain of about 100 is used, the simulated bandwidth for differential-mode input currents is 2.85 GHz. Common-mode input currents experience a DC transimpedance of only 44.8 dBΩ, as can also be seen in Fig. 4.36; this means that the value of $R_{out,CM}$ of the last amplifier stage equals 174 Ω = 44.8 dBΩ (4.119). The bandwidth for common-mode signals is about 100 times lower than for differential-mode signals, which is also as expected. The simulated RMS input noise current of this TIA is 624 nA_{RMS} of which 119 nA_{RMS} is produced by the two feedback resistors and 619 nA_{RMS} comes from the differential voltage amplifier.

If the second voltage amplifier design is used and the feedback resistance is unaltered, the DC transimpedance is the same as before, while the bandwidth decreases slightly to 2.71 GHz. This is due to the lower frequency of the pole that is located at the output of the first stage which now has a higher output resistance due to its different biasing condition. The behavior for common-mode signals is the same as when the other voltage amplifier is used. The Bode plot of this TIA is not shown because it is almost identical to that shown in Fig. 4.36. The simulated RMS input noise current of this TIA is 669 nA_{RMS} of which 116 nA_{RMS} comes from the two

[15] This parasitic capacitance is seen at both terminals of the photodiode because the illuminated and non-illuminated photodiodes have the same physical size.

Table 4.4 Comparison of the two presented differential shunt-shunt feedback TIAs

	Design 1	Design 2
$\lvert Z_{TIA,DM,DC} \rvert$	4.95 kΩ	4.95 kΩ
$\lvert Z_{TIA,CM,DC} \rvert$	174 Ω	174 Ω
BW_{TIA}	2.85 GHz	2.71 GHz
input noise	624 nA_{RMS}	669 nA_{RMS}
feedback resistors	119 nA_{RMS}	116 nA_{RMS}
voltage amplifier	619 nA_{RMS}	659 nA_{RMS}
current consumption	77.4 mA	31.5 mA

feedback transistors and 659 nA_{RMS} is caused by the differential voltage amplifier. Concluding, the RMS input noise current of the second TIA is only 7% higher than that of the first TIA that consumes 146% more current.

In the frequency range of interest, namely between about 1 MHz and the bandwidth, the transimpedance for common-mode signals is at least 30 dB lower than for differential-mode signals which is enough to ensure proper subtraction of the photodiode currents. The performance of both designs is summarized in Table 4.4.

4.4.2 A Shunt-Shunt Feedback TIA for POF-Applications

If a POF is used instead of a MMF, the area of the photodiode should increase significantly due to the large diameter of this type of fiber, namely 1 mm typically. This results in a massive photodiode capacitance of 63.6 pF for an n-well/p-sub photodiode with a diameter of 1 mm in a 180 nm CMOS process and a reverse voltage of 0.9 V which is half of the nominal supply voltage of the technology. The circuit diagram of the single-ended shunt-shunt feedback TIA that is designed to operate with this large-area photodiode is shown in Fig. 4.37. Three different building blocks can be distinguished: the TIA itself, the output buffer, and the DC current compensation circuit.

The voltage amplifier of the TIA is composed of three cascaded inverting amplifying stages, as can be seen in Fig. 4.37. Whereas the first stage is a double common-source amplifier, the subsequent two stages are common-source stages. A double common-source amplifier is used at the input because such a stage is able to provide significantly more gain than a common-source amplifier; this property is advantageous to suppress the noise of the subsequent stages. The high input capacitance of such an amplifier stage does not pose a problem here because the input capacitance of this TIA is nevertheless dominated by the large photodiode capacitance of 63.6 pF. The second and third amplifying stages are common-source stages with a significantly smaller input capacitance. These prevent the poles on the intermediate circuit nodes from shifting to too low a frequency, which would jeopardize the stability of the system. Instead of using resistors as loads for the common-source amplifiers, as described in Sect. 4.3.1, active loads are used here. To guarantee correct biasing of these active loads, which are implemented as pMOS transistors in their saturation region, their biasing voltages are derived from the signal path by

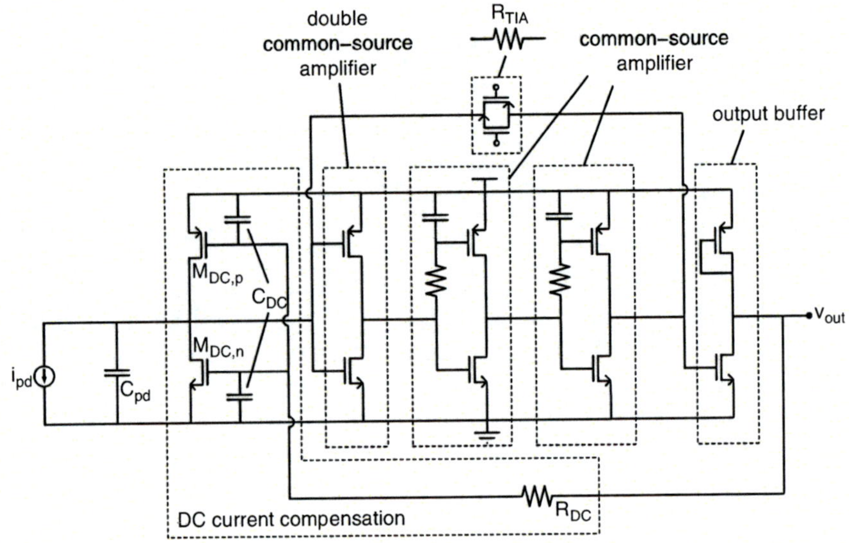

Fig. 4.37 Circuit diagram of the single-ended TIA with DC current compensation to be used with a large-area integrated photodiode in 180 nm CMOS

means of low-pass filters, as can be seen in Fig. 4.37. For frequencies below the cut-off frequency of the low-pass filters, the common-source amplifiers actually act as double common-source amplifiers with a correspondingly high gain and large input capacitance. Above the cut-off frequency, the large input capacitance of the pMOS is removed from the signal path and the voltage gain of the amplifiers decrease as a result. The feedback resistance of this shunt-shunt feedback TIA is implemented as a parallel combination of an nMOS and a pMOS in their linear region. Because the inherent offset of the photodiode current[16] is eliminated by means of the DC current compensation block discussed below, both an nMOS and a pMOS are used in order to obtain symmetrical behavior for currents flowing into or out of the TIA.

The output buffer in Fig. 4.37 is used to lower the output impedance of the TIA. More importantly, however, it generates a voltage with the correct polarity to be used in the DC current compensation block. It is implemented as a common-source stage of which the active load is biased as a diode. This again guarantees that the biasing of this stage is the same as that of the three amplifying stages so that the DC voltage through the complete circuit is identical on all signal nodes.

A separate path for the DC current of the integrated photodiode is provided by means of an extra feedback path, as can be seen in Fig. 4.37. This is done in order

[16] When illuminated, an n-well/p-sub photodiode generates a current that flows out of the TIA while there is no current when the photodiode in not illuminated. The average photodiode current is consequently not positioned around 0 A.

not to disturb the correct biasing of the transistors, especially those in the last amplification stage where the voltage swing is the largest. For frequencies below the cut-off frequency of the low-pass filter R_{DC}-C_{DC}, the transconductance of transistors $M_{DC,n}$ and $M_{DC,p}$ appears in parallel with the regular feedback path with equivalent resistance R_{TIA}. The transimpedance at these low frequencies, $Z_{TIA,lf}$, can then be calculated as follows:

$$
\begin{aligned}
Z_{TIA,lf} &= R_{TIA} // \frac{1}{g_{m,M_{DC,n}} + g_{m,M_{DC,p}}} \\
&\approx \frac{1}{g_{m,M_{DC,n}} + g_{m,M_{DC,p}}}.
\end{aligned} \tag{4.122}
$$

At frequencies above the cut-off frequency of this low-pass filter, the transimpedance is simply determined by the value of R_{TIA}, like in a regular shunt-shunt feedback TIA. The two extra transistors that are needed to construct the DC current compensation path, namely $M_{DC,n}$ and $M_{DC,p}$, also contribute to the overall noise of this TIA. The current noise of these transistors can simply be added to the photodiode current because they are in parallel. Since the spectral density of the channel noise of a transistor is proportional to its transconductance (4.27), the latter should be minimized to minimize the noise of the TIA. Consequently, there is a trade-off between the noise performance and the DC current compensation capability because the latter is proportional to the transconductance of these transistors (4.122).

The equivalent resistance of the feedback network consisting of the parallel combination of an nMOS and a pMOS in their linear region is chosen approximately equal to 58 kΩ, which results in a simulated transimpedance of 95.1 dBΩ = 56.9 kΩ. As already mentioned, the photodiode capacitance has a value of 63.6 pF. Due to an overdrive voltage of only 50 mV for all transistors in the voltage amplifier[17], the gain of its first stage has an impressively high simulated value of 27.7 = 28.8 dB while that of the second and third stages equal 12.7 = 22.1 dB. These values result in a simulated bandwidth of 139 MHz, as can be seen in the Bode plot in Fig. 4.38. The output buffer has a simulated gain of −0.913 dB = 0.900. The low-pass filters that take care of the biasing of the pMOS loads in the second and third amplifying stage have a cut-off frequency of 232 kHz. By contrast, the cut-off frequency of the low-pass filter R_{DC}-C_{DC} is chosen equal to 250 Hz, resulting in a cut-off frequency of 8.76 kHz, as can be verified in Fig. 4.38a. The transimpedance for frequencies below this cut-off frequency is equal to 1.49 kΩ which is determined by the transconductances of $M_{DC,n}$ and $M_{DC,p}$, both conducting a current of 21.0 μA, as predicted by (4.122). Consequently, the DC current of the photodiode experiences an amplification which is approximately 40 times lower than that of the signal components in the frequency range of interest. To optimize the noise performance of this TIA, the major portion of the total current, namely 12.0 mA, is reserved for the first amplifying stage, while the second and third stages conduct currents of 1.42 mA and 177 μA, respectively. The output buffer of Fig. 4.37 consumes a current of 88.7 μA.

[17] This is realized by deliberately choosing the supply voltage equal to 1.12 V.

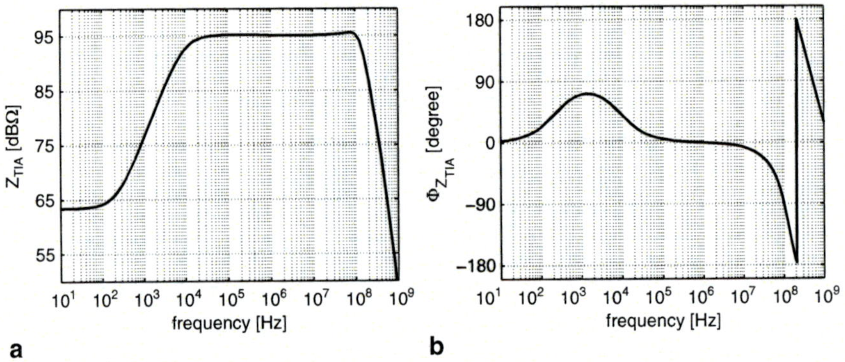

Fig. 4.38 Simulated Bode plot of the implemented single-ended shunt-shunt feedback TIA with a 1 mm integrated photodiode in 180 nm CMOS. **a** Magnitude. **b** Phase

Table 4.5 Overview of the simulated results of the single-ended shunt-shunt feedback TIA with a 1 mm integrated photodiode in 180 nm CMOS

$\lvert Z_{TIA,lf} \rvert$	1.49 kΩ
$\lvert Z_{TIA,hf} \rvert$	56.9 kΩ
BW_{TIA}	139 MHz
input noise	137 nA_{RMS}
current consumption	13.7 mA

This results in an RMS input noise current of the complete TIA of 137 nA_{RMS}. The performance figures of this design are summarized in Table 4.5.

4.4.3 A Capacitance-Relieved TIA for POF-Applications

The purpose of this design is to show the benefits of the capacitance-relieved TIA compared to a regular shunt-shunt feedback TIA when a large-area photodiode is used. For an integrated n-well/p-sub photodiode in the used 130 nm CMOS technology with a diameter of 1 mm and a reverse voltage of 0.6 V, corresponding to half of the nominal supply voltage, this leads to a photodiode capacitance of 106 pF. The capacitance-relieved TIA is therefore an attractive topology for this application because it can combine a large transimpedance and a high bandwidth even for a large input capacitance.

The schematic of the proposed design is shown in Fig. 4.39. By matching the block diagram of the capacitance-relieved TIA in Fig. 4.30b to the schematic in Fig. 4.39, it can be seen that the first voltage amplifier with gain A is provided by a cascade of two inverting common-source stages, while the second voltage amplifier with gain B is provided by a third common-source stage. In Sect. 4.3.3, it has been assumed that $B = 1$. As B is realized here by a common-source amplifier with a significantly higher gain, the formulas derived in Sect. 4.3.3 cannot be used. Nevertheless, the

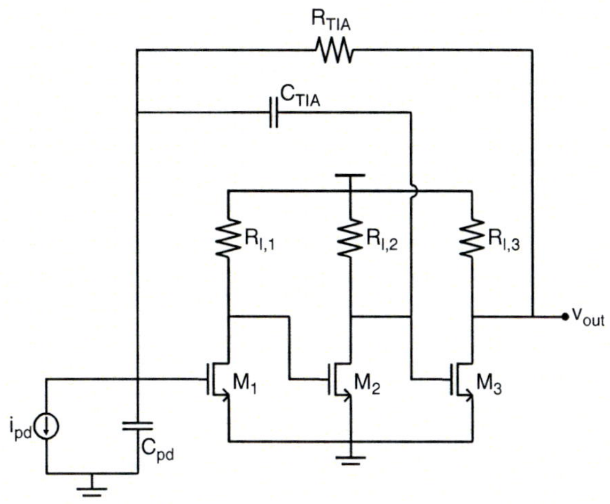

Fig. 4.39 Circuit diagram of the capacitance-relieved TIA with a large-area integrated photodiode in 130 nm CMOS

DC transimpedance is still determined by the value of R_{TIA}. The bandwidth of the presented design can be deduced from (4.103) and is given by the following formula:

$$BW = \frac{1}{2\pi R_{TIA}\left(\frac{C_{pd}}{AB} - \frac{C_{TIA}}{B}\right)}. \tag{4.123}$$

All three common-source amplifier stages in Fig. 4.39 have equal sizes and bias conditions. To achieve good power efficiency, the overdrive voltages of M_1, M_2 and M_3 are chosen equal to 50 mV, as explained in Sect. 4.4.1. The transistors M_1, M_2 and M_3 are 120 nm long, the minimum length of the technology used and 250 µm wide. The drain-source current through each of the transistors then equals 3.32 mA, resulting in a transconductance of 57.2 mS, as predicted by Fig. 4.20a. With an intrinsic transistor gain of 12.9 for the chosen overdrive voltage (Fig. 4.20b), the small-signal output resistances of M_1, M_2 and M_3 equal 226 Ω. The value of the load resistors R_1, R_2 and R_3 must be 240 Ω to ensure the correct bias condition for the three transistors. The total output resistance, which is the parallel combination of the load resistance and the transistor output resistance, is therefore equal to 116 Ω. The voltage gain of one stage can now be calculated to be 6.64. The voltage gains A and B thus equal 44.1 and 6.64, respectively. R_{TIA} is fixed at 100 kΩ. The total power consumption is 12.0 mW if the supply voltage is 1.2 V, the nominal supply voltage of the technology.

Without capacitor C_{TIA} in Fig. 4.39, the circuit is actually a regular shunt-shunt feedback TIA. The bandwidth can then be calculated with (4.45) and equals 4.40 MHz. If a feedback capacitor C_{TIA} with a value of 2.08 pF is attached, the bandwidth of the TIA increases to 40.3 MHz, which is slightly higher than what is predicted by (4.123); this is due to the proximity of the non-dominant pole at the output node of the second amplifier stage. The bandwidth is thus increased by a factor of 9 without any supplementary power consumption. The same bandwidth could also be

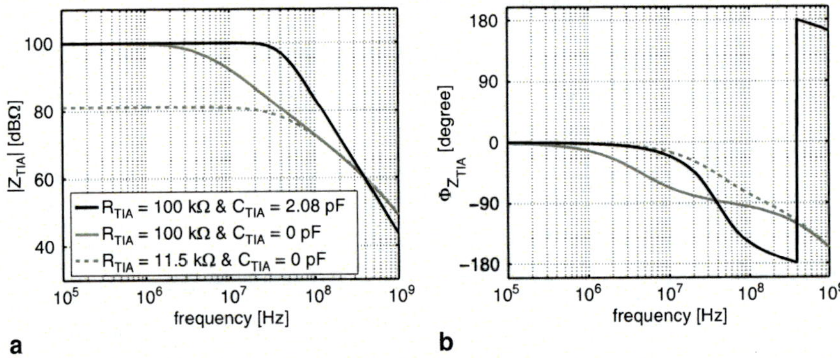

Fig. 4.40 Simulated Bode plot of the implemented capacitance-relieved TIA. **a** Magnitude. **b** Phase

Table 4.6 Overview of the three design cases of the capacitance-relieved TIA

R_{TIA}	C_{TIA}	BW_{TIA}	RMS input noise
100 kΩ	0 pF	4.40 MHz	44.4 nA_{RMS}
100 kΩ	2.08 pF	40.3 MHz	90.3 nA_{RMS}
10.9 kΩ	0 pF	40.3 MHz	399 nA_{RMS}

obtained without capacitor C_{TIA} by lowering the value of R_{TIA} to 10.9 kΩ. However, the DC transimpedance would decrease accordingly. The simulated Bode plots of all three mentioned cases are shown in Fig. 4.40 and the corresponding bandwidths are summarized in Table 4.6.

The noise of the receiver is determined by the transconductance of M_1 and by the value of the feedback resistor R_{TIA}, as can be seen in (4.111). To enable a fair comparison between all three cases presented, the simulated integrated output noise is divided by the DC transimpedance to obtain the equivalent integrated input noise. These values are listed in Table 4.6. The addition of C_{TIA} with a value of 2.08 pF results in an increase in the bandwidth by a factor of 9 and in an increase in the noise level from 44.4 to 90.3 nA_{RMS}. The RMS input noise current when R_{TIA} is only 11.5 kΩ and the feedback capacitor is removed is to 399 nA_{RMS}. Thus, for an equal bandwidth, the sensitivity of the receiver with capacitor C_{TIA} attached is 4.4 times better compared to the case when that bandwidth is obtained by reducing the value of the feedback resistor R_{TIA}. The fact that this rise in performance is achieved without any additional power consumption demonstrates the efficiency of the capacitance-relieved TIA.

4.5 Conclusion

After being converted to the electrical domain by the photodiode, the signal needs to be transformed into a voltage; this is realized by the TIA. This building block has been treated extensively in this chapter.

The key specifications of a TIA are discussed first. It is shown that the frequency dependent transimpedance contains two mutually dependent parameters, namely the DC transimpedance and the bandwidth. These are combined in the transimpedance-bandwidth product, a figure which depends primarily on the photodiode capacitance and the chosen TIA topology. The circuit noise of the TIA is another major performance measure, especially when taking into account the low responsivity of an integrated CMOS photodiode. The TIA overload current is less important, for the same reason.

An overview of the possible topologies for a TIA is then given. Both open-loop topologies, such as the single-resistor TIA and the common-gate TIA, and closed-loop topologies, such as the regulated-cascode TIA and the shunt-shunt feedback TIA, are compared to each other in terms of transimpedance gain, bandwidth, noise and power consumption. It is shown that the shunt-shunt feedback TIA is able to combine high transimpedance gain, high bandwidth and good low-noise performance. It is therefore applied throughout this work.

Some design aspects of the shunt-shunt feedback TIA are discussed next. The voltage amplifier can be a simple single-stage voltage amplifier or a cascade of several such stages. By applying the downscaling principle, it is shown that the trade-off between bandwidth, noise and power consumption can be improved significantly. The feedback network can best be implemented by a transistor in its linear region to circumvent the inherent problems of polysilicon resistors with a large resistance. The capacitance-relieved TIA is proposed to improve the trade-off between transimpedance gain and bandwidth in a shunt-shunt feedback TIA.

To conclude, three designs with their simulation results are discussed.

Chapter 5
Increasing the Speed—The Equalizer

The voltage signal at the output of the TIA is large enough to dominate the noise of the subsequent building blocks. However, the major problem at this point in the receiver chain is the insufficient bandwidth of the signal which is due to the very limited speed of the integrated photodiode in CMOS. The equalizer tackles this issue.

The operating principle of the equalizer is discussed first. It is shown how the equalizer should behave so that the limited bandwidth of the integrated photodiode is compensated for. It is demonstrated that the appropriate characteristic can be realized by placing several first order filter branches in parallel.

As the equalizer is implemented here as an analog building block, two circuit techniques are proposed thereafter that realize a single filter branch: source degeneration and inductive loading.

Finally, two circuit designs are presented to end this chapter.

5.1 Operation Principle

It has been shown in Chap. 3 that the frequency response of an n-well/p-sub photodiode which is integrated in a contemporary CMOS technology has a frequency response that is not suitable for high-speed applications. It has been derived theoretically in Sect. 3.2.1 that the -3 dB frequency of such a device, when illuminated with 850 nm light, is no greater than about 2.5 MHz; this was also found in [Gen01, Rad06]. Above this -3 dB frequency, the magnitude of the photodiode response rolls-off at a rate between 4 dB/decade and 10 dB/decade as shown in Chap. 3. This is significantly slower than the roll-off of an electrical first order low-pass filter which decays at a rate of 20 dB/decade above its corner frequency. This leads to the very important observation that, thanks to this relatively slow decay, the frequency components above the bandwidth of the photodiode are only moderately attenuated. This opens the path to equalization because a signal with a relatively high SNR can be present, even for frequencies in the GHz-range [Rad06].

The working principle of the photodiode equalizer is shown in Fig. 5.1. As the equalizer is placed in the signal path and consequently in series with the photodiode,

F. Tavernier, M. Steyaert, *High-Speed Optical Receivers with Integrated Photodiode in Nanoscale CMOS,* Analog Circuits and Signal Processing 5,
DOI 10.1007/978-1-4419-9925-2_5,

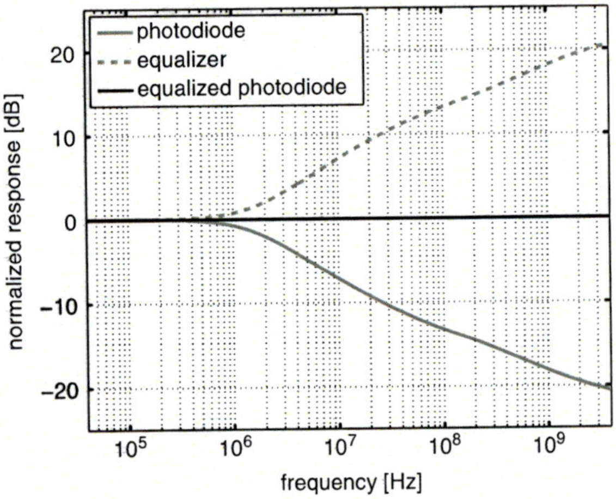

Fig. 5.1 Working principle of the photodiode equalizer

both responses should be multiplied to obtain the combined response[1]. The equalizer should have a magnitude response which is the inverse of the magnitude response of the integrated photodiode in order to obtain a combined behavior with a constant magnitude response at all frequencies. This ensures that all the frequency components of the received light pulse experience the same amplification. The same has to be done regarding the phase response of the photodiode. In order to obtain a flat phase response in the frequency domain, the equalizer should realize the opposite of the phase response of the integrated photodiode. Although this photodiode property has not been discussed in Chap. 3, it is treated in [Rad06] for a regular n-well/p-sub photodiode and in [Her07] for a regular and a differential n-well/p-sub photodiode. In harmony with the magnitude having a decay which is slower than that of a first order low-pass filter, the phase rotation above the photodiode −3 dB frequency is smaller than −90°. In [Rad06], it is confirmed that the phase shift of a regular n-well/p-sub photodiode is −10° at the −3 dB frequency, −15° between 10 and 100 MHz and gradually increases to −20° for higher frequencies. In [Her07], it is concluded that the phase shift of a differential n-well/p-sub photodiode is approximately −20° at the −3 dB frequency and increases for higher frequencies.

Realizing an electrical circuit with the same low roll-up as the roll-off of the integrated photodiode is not straightforward because the roll-up of a first order high-pass filter has a slope of 20 dB/decade, which is too steep. It has been proposed in [Rad05] to realize such a response by placing several first order high-pass filters in parallel with a direct branch. This is visualized in Fig. 5.2 where it is assumed that the equalization is performed in the voltage domain. The transfer function of such

[1] The multiplication is transformed into addition if the quantities are expressed on a logarithmic scale.

Fig. 5.2 Block diagram of an equalizer with n high-pass filters in parallel with a direct branch

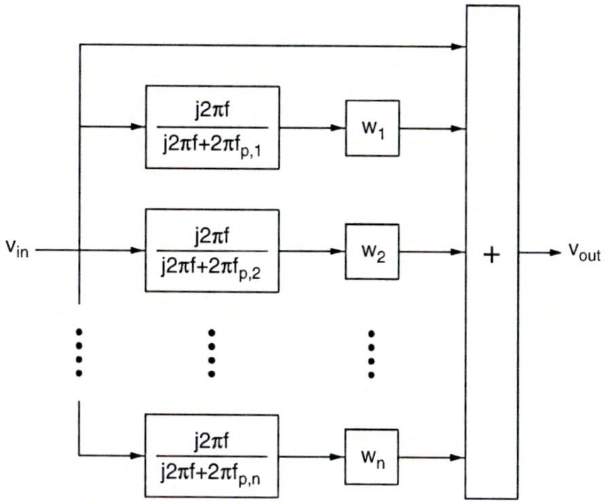

an equalizer can be written as follows:

$$E = 1 + w_1 \cdot \frac{j2\pi f}{j2\pi f + 2\pi f_{p,1}} + w_2 \cdot \frac{j2\pi f}{j2\pi f + 2\pi f_{p,2}} + \cdots + w_n \cdot \frac{j2\pi f}{j2\pi f + 2\pi f_{p,n}}, \tag{5.1}$$

where w_i and $f_{p,i}$ represent the weight and corner frequency of the ith high-pass filter, respectively. An equalizer with a low roll-up is obtained by a deliberate choice of these parameters. The Bode plot of such an equalizer with four parallel high-pass filters is shown in Fig. 5.3. It can be seen clearly in Fig. 5.3a that the low roll-up is

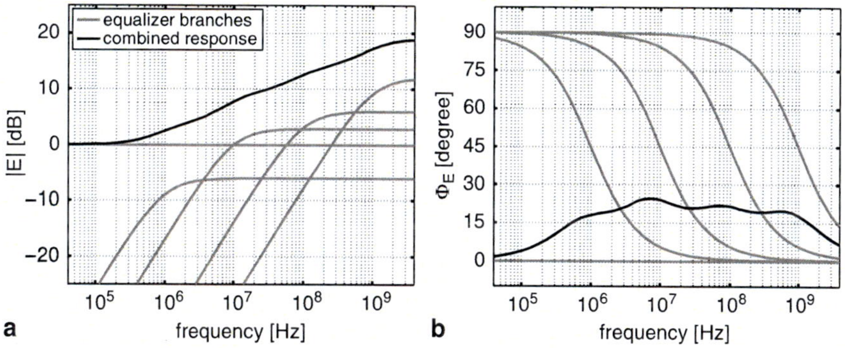

Fig. 5.3 Bode plot of the equalizer which is realized by adding the frequency responses of four high-pass filters with prescribed weights and corner frequencies to a direct path between the input and output. **a** The magnitude of the equalizer is determined by the gain boost provided by every filter branch. **b** The phase of the equalizer is determined by the phase boost provided by every filter branch

not realized by the 20 dB/decade slope of the high-pass filters. Rather, it is obtained by the behavior of the filters around their respective corner frequencies where the magnitude has a slope that evolves gradually from 20 to 0 dB/decade. The phase response of such an equalizer as visualized in Fig. 5.3b shows a rotation of maximally 25° which is approximately equal to the negative phase of an integrated photodiode, as discussed before. It can therefore be concluded that an equalizer with a topology as shown in Fig. 5.2 can equalize both the magnitude and the phase response of an integrated regular or differential n-well/p-sub photodiode.

At this point, the problem of realizing an equalizer with an appropriate frequency response can be split in two parts:

1. How many high-pass filters are needed?
2. What should be the weight w_i and the corner frequency $f_{p,i}$ of each high-pass filter?

This problem can only be solved numerically by an optimization algorithm because there is no mathematical expression which relates the frequency response of the photodiode to the weights and corner frequencies of the high-pass filters. The outcome of this optimization algorithm depends strongly on the type of photodiode that is used and the frequency range over which the compensation needs to be performed. Generally it can be stated that more high-pass filters are needed to equalize a regular n-well/p-sub photodiode than to equalize a differential n-well/p-sub photodiode which is thanks to the intrinsically higher bandwidth of the latter device. In addition to this, it can easily be understood that more filter branches are needed if the photodiode response needs to be compensated over a wider frequency range.

Most of the implementations that are presented in Chap. 7 make use of a differential n-well/p-sub photodiode. Therefore, the discussion here is concentrated towards the equalization of this kind of photodiodes. It is also assumed that the equalization needs to be performed up to a frequency of 5 GHz. The applied optimization algorithm, which is a genetic algorithm[2], resulted in the conclusion that an equalizer with two high-pass filters can adequately equalize a differential n-well/p-sub photodiode up to 5 GHz. The optimized parameters of this equalizer are summarized in Table 5.1. The Bode plot of this equalizer is shown in Fig. 5.4.

Table 5.1 Parameters of the high-pass filters of an equalizer with two filter branches which is optimized to equalize the response of a differential n-well/p-sub photodiode

w_1	0.339 = −9.40 dB
$f_{p,1}$	398 MHz
w_2	1.48 = 3.41 dB
$f_{p,2}$	5.43 GHz

[2] A genetic algorithm aims at emulating the process of natural evolution to find the optimal solution of a problem.

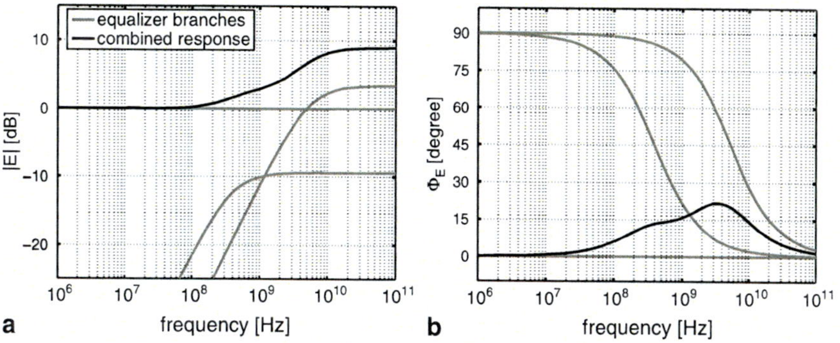

Fig. 5.4 Bode plot of the optimized equalizer consisting of a direct branch and two high-pass filters with the parameters of Table 5.1. **a** Magnitude. **b** Phase

5.2 Circuit Techniques

A first order high-pass filter can be realized in several ways. The most straightforward possibility is to implement it passively using a resistor and a capacitor or a resistor and an inductor. The response of every filter branch should also be amplified or attenuated by a dedicated weight. This task cannot be performed by a simple passive filter. An amplifier is therefore needed for every filter branch in the equalizer. In addition, the signals of all filter branches must be added, which again requires an extra circuit. In the following, it will be shown how such a combination of a passive filter, an amplifier and a circuit taking care of the addition can be merged into a single amplifier stage.

5.2.1 A Source-Degenerated Amplifier

A first possibility to integrate the functionality of the filter and the amplifier is shown in Fig. 5.5a. Due to the presence of the current source at the source of transistor M_e, no small-signal current can flow through the transistor at low frequencies and the voltage gain between v_{in} and v_{out} is consequently 0. However, a small-signal current can flow through R_1 at very high frequencies because capacitor C_1 then behaves as a short circuit. As a result, the voltage gain of this stage is larger than 0 at these high frequencies. A corner frequency is therefore established somewhere in between. This behavior can also be deduced from the transfer function of the presented stage:

$$A_{sd} = -\frac{j2\pi f R_l C_1}{1 + j2\pi f \left(R_1 + \frac{1}{g_{m,M_e}}\right) C_1}. \tag{5.2}$$

When the inverting character is ignored, this transfer function can be translated into a weight w_1 and a corner frequency $f_{p,1}$ of a filter branch in Fig. 5.2 by the following equalities:

$$w_1 = \frac{R_l}{R_1 + \frac{1}{g_{m,M_e}}}, \tag{5.3}$$

$$f_{p,1} = \frac{1}{2\pi \left(R_1 + \frac{1}{g_{m,M_e}}\right) C_1}. \tag{5.4}$$

If it is assumed that the transconductance of M_e is large compared to the conductance of R_1, the weight of this high-pass filter is set by the ratio of R_l and R_1, while its corner frequency is determined only by R_1 and C_1.

It is possible to combine two or more high-pass filters with their respective weights and corner frequencies into a single stage simply by adding parallel branches at the source of transistor M_e, each consisting of a resistor R_i and a capacitor C_i. If, in addition, the current source in Fig. 5.5a is replaced by a resistor, the unfiltered branch of Fig. 5.2 can also be realized by this stage. This is shown in Fig. 5.5b for an equalizer with a direct branch and two high-pass filters. If the transconductance of transistor M_e is again assumed to be much larger than the conductance of resistors R_0, R_1 and R_2, the weights of the filter branches in Fig. 5.5b can be calculated as follows:

$$w_0 = \frac{R_l}{R_0}, \tag{5.5}$$

$$w_1 = \frac{R_l}{R_1}, \tag{5.6}$$

$$w_2 = \frac{R_l}{R_2}, \tag{5.7}$$

where w_0 represents the weight of the unfiltered branch of the equalizer. Until now, this weight has been assumed to be equal to 1 which requires equal values for R_l

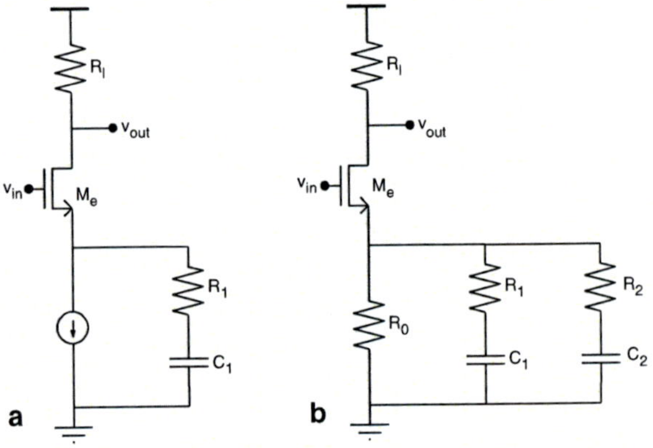

Fig. 5.5 Circuit diagram of an active high-pass filter realized by means of frequency dependent source degeneration. **a** Implementation of a single high-pass filter. **b** Implementation of an equalizer with two filter branches

and R_0. The corner frequencies of the two high-pass filters that are realized by the circuit in Fig. 5.5b are defined by the following equations:

$$f_{p,1} = \frac{1}{2\pi R_1 C_1}, \quad (5.8)$$

$$f_{p,2} = \frac{1}{2\pi R_2 C_2}. \quad (5.9)$$

If the weight and corner frequency of each filter branch in Fig. 5.2 is known, the resistances R_i and capacitances C_i can be calculated with the help of the formulas above. Note that the implementation of Fig. 5.5b provides an extra degree of freedom, for example the value of the load resistor R_l. This load resistance should be maximized in order to minimize the power consumption and the capacitances C_i. By contrast, a large load resistance leads to a pole at the output node of the equalizer which can be located at too low a frequency.

5.2.2 A Common-Source Amplifier with an Inductive Load

The impedance at the source of the input transistor is modified as a function of frequency in the source-degenerated amplifier. By contrast, the common-source amplifier with an inductive load modifies the impedance at the drain of the same input transistor to obtain a comparable result. Its topology is shown in Fig. 5.6a. The inductor L_1 shorts the resistor R_1 at low frequencies, which results in a voltage gain of 0 between v_{in} and v_{out}. At higher frequencies, however, the effect of the inductor disappears and the voltage gain increases accordingly. This behavior can be verified in the transfer function of this stage:

$$A_{il} = -\frac{j2\pi f g_{m,M_e} L_1}{1 + j2\pi f \frac{L_1}{R_1}}. \quad (5.10)$$

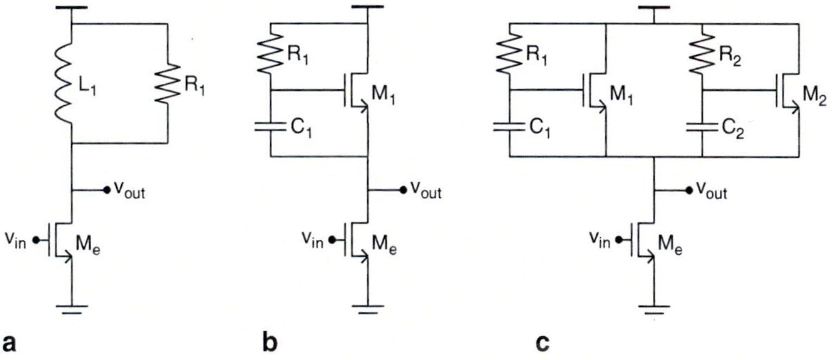

Fig. 5.6 Circuit diagram of an active high-pass filter realized by means of an inductive load network. **a** Passive inductor. **b** Active inductor. **c** Implementation of an equalizer with two filter branches

If the inverting character is ignored, this transfer function can be translated into a weight w_1 and a corner frequency $f_{p,1}$ of an active high-pass filter which are defined by the following relations:

$$w_1 = g_{m,M_e} R_1, \tag{5.11}$$

$$f_{p,1} = \frac{1}{2\pi \frac{L_1}{R_1}}. \tag{5.12}$$

The expression for the weight is clearly the same as that of the gain of a normal common-source amplifier because the effect of the inductor can be neglected at high frequencies.

Inductors are difficult to integrate on a chip. They occupy a large area, which has severe consequences for the cost and they suffer from a low quality factor due to the presence of a conducting substrate. It is therefore of interest to emulate inductive behavior by means of devices that are easier to integrate, namely transistors, resistors and capacitors. This can be done in this case by modifying the load network, as shown in Fig. 5.6b. The small-signal load impedance at low frequencies is determined by the transconductance of transistor M_1, which also determines the gain between v_{in} and v_{out}. The gate and source of M_1 are shorted by capacitor C_1 at high frequencies and therefore no small-signal current can flow through M_1. At these frequencies, the load impedance is determined by R_1. If the conductance of R_1 is significantly lower than the transconductance of M_1, the voltage gain at high frequencies is significantly higher than at low frequencies. This is equivalent to the inductive behavior of the circuit in Fig. 5.6a. The transfer function of the circuit in Fig. 5.6b is given by the following formula:

$$A_{ail} = -\frac{g_{m,M_e}(1 + j2\pi f R_1 C_1)}{g_{m,M_1} + j2\pi f C_1}. \tag{5.13}$$

Instead of having a zero at DC, this stage has a zero at a frequency which is determined by the time constant of R_1 and C_1. It can therefore only be considered as a high-pass filter above this frequency. Neglecting the minus sign in (5.13), the weight and corner frequency of the high-pass filter can be calculated as follows:

$$w_1 = g_{m,M_e} R_1, \tag{5.14}$$

$$f_{p,1} = \frac{g_{m,M_1}}{2\pi C_1}. \tag{5.15}$$

Apart from the fact that this stage only acts as a high-pass filter at frequencies above the zero frequency in (5.13), there is another major difference between the circuit with the real inductor in Fig. 5.6a and that with the active inductive load, namely the noise performance. As the effect of the inductor, which is a noiseless device, is emulated by a transistor with a non-zero transconductance, the noise performance of the circuit with the active inductor is always worse than when a real inductor is used. However, this is not a significant issue here because the equalizer is located after the

TIA, which guarantees that the signal has a relatively large swing at the equalizer input.

Multiple active inductors can be placed in parallel to realize a complete equalizer in a single stage. This is shown in Fig. 5.6c for an equalizer with two active inductors implementing two high-pass filters. The property of the active inductor to provide voltage gain at low frequencies is used here to implement the direct branch of the equalizer. The weights and corner frequencies of the two high-pass filters cannot readily be deduced however. The transfer function of the equalizer in Fig. 5.6c is given by:

$$\begin{aligned}\frac{v_{out}}{v_{in}} &= -\frac{g_{m,M_e}}{g_{m,M_1}+g_{m,M_2}}\\ &\times \frac{1+j2\pi f\left(R_1C_1+R_2C_2\right)-4\pi^2 f^2 R_1^2R_2^2C_1^2C_2^2}{1+j2\pi f\frac{\left(1+g_{m,M_2}R_1\right)C_1+\left(1+g_{m,M_1}R_2\right)C_2}{g_{m,M_1}+g_{m,M_2}}-4\pi^2 f^2\frac{\left(R_1+R_2\right)C_1C_2}{g_{m,M_1}+g_{m,M_2}}}\end{aligned} \tag{5.16}$$

As an equalizer with two high-pass filters is realized by this transfer function, the following relation should also be valid:

$$\begin{aligned}\frac{v_{out}}{v_{in}} &= w_0 + w_1\cdot\frac{j2\pi f}{j2\pi f+2\pi f_{p,1}} + w_2\cdot\frac{j2\pi f}{j2\pi f+2\pi f_{p,2}}\\ &= w_0\cdot\frac{1+jf\left(\frac{1}{f_{p,1}}\left(1+\frac{w_1}{w_0}\right)+\frac{1}{f_{p,2}}\left(1+\frac{w_2}{w_0}\right)\right)-\frac{f^2}{f_{p,1}f_{p,2}}\left(1+\frac{w_1}{w_0}+\frac{w_2}{w_0}\right)}{1+jf\left(\frac{1}{f_{p,1}}+\frac{1}{f_{p,2}}\right)-\frac{f^2}{f_{p,1}f_{p,2}}}.\end{aligned} \tag{5.17}$$

To implement the same equalizer characteristic, the respective terms in (5.16) and (5.17) should be equal to each other. This leads to the following set of equations:

$$\begin{cases} w_0 = \dfrac{g_{m,M_e}}{g_{m,M_1}+g_{m,M_2}}\\[2ex] \dfrac{1}{f_{p,1}}\left(1+\dfrac{w_1}{w_0}\right)+\dfrac{1}{f_{p,2}}\left(1+\dfrac{w_2}{w_0}\right) = 2\pi\left(R_1C_1+R_2C_2\right)\\[2ex] \dfrac{1}{f_{p,1}f_{p,2}}\left(1+\dfrac{w_1}{w_0}+\dfrac{w_2}{w_0}\right) = 4\pi^2R_1^2R_2^2C_1^2C_2^2\\[2ex] \dfrac{1}{f_{p,1}}+\dfrac{1}{f_{p,2}} = 2\pi\dfrac{\left(1+g_{m,M_2}R_1\right)C_1+\left(1+g_{m,M_1}R_2\right)C_2}{g_{m,M_1}+g_{m,M_2}}\\[2ex] \dfrac{1}{f_{p,1}f_{p,2}} = 4\pi^2\dfrac{\left(R_1+R_2\right)C_1C_2}{g_{m,M_1}+g_{m,M_2}} \end{cases} \tag{5.18}$$

To simplify this non-linear set of equations, it is assumed here that the zeros and poles in (5.16) and (5.17) are located far from each other ($f_{p,1} < f_{p,2}$). The solution of (5.18) is then as follows:

$$\begin{cases} w_0 = \dfrac{g_{m,M_e}}{g_{m,M_1} + g_{m,M_2}} \\ w_1 = \dfrac{g_{m,M_e}}{g_{m,M_1} + g_{m,M_2}} \cdot \dfrac{g_{m,M_1} R_1 - 1}{g_{m,M_2} R_1 + 1} \\ w_2 = g_{m,M_e} \cdot \dfrac{R_1^2}{R_1 + R_2} \cdot \dfrac{g_{m,M_2} R_2 - 1}{g_{m,M_2} R_1 + 1} \\ f_{p,1} = \dfrac{g_{m,M_1} + g_{m,M_2}}{2\pi \left(1 + g_{m,M_2} R_1\right) C_1} \\ f_{p,2} = \dfrac{1 + g_{m,M_2} R_1}{2\pi \left(R_1 + R_2\right) C_2} \end{cases} \tag{5.19}$$

It can be concluded from this solution that the voltage gain at low frequencies, w_0, is determined by the transconductances of M_e, M_1 and M_2. As the current through M_e equals the sum of the currents through M_1 and M_2, w_0 is equal to 1 if all transistors have the same overdrive voltage. The weight of the first high-pass filter can be set by the ratio of g_{m,M_1} and g_{m,M_2} and the value of R_1. The value of R_2 then determines the value of w_2. With all the weights fixed, the corner frequency of the first high-pass filter is set by C_1 while that of the second high-pass filter is set by C_2.

5.3 Equalizer Designs

To illustrate the circuit concepts that have been introduced in the previous section, two designs and their simulation results are shown in this section. The first design realizes its frequency response by means of frequency dependent source degeneration while the second design applies the principle of active inductors.

5.3.1 *A Differential Equalizer with Differential Source Degeneration*

This equalizer is intended to compensate the response of a differential n-well/p-sub photodiode in a 130 nm CMOS technology. In Chap. 3, it has been shown that the −3 dB bandwidth of such a device is between 377 MHz and 1.47 GHz for a periodicity which varies between 10 and 5 μm. Although these values are more than

Fig. 5.7 Circuit diagram of the differential equalizer with differential source degeneration and common-mode rejection circuit in 130 nm CMOS

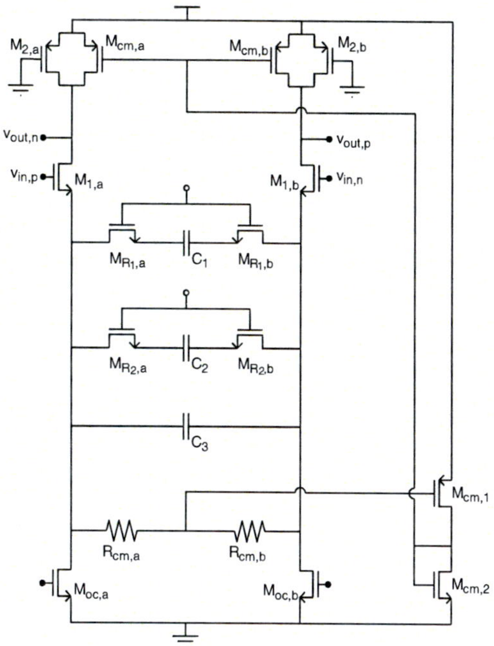

100 times greater than the bandwidth of a regular n-well/p-sub photodiode in the same technology, they are too low to permit multi-Gbit/s communication.

The circuit diagram of the equalizer is shown in Fig. 5.7. It is designed to operate with differential-mode input signals that are generated by the differential TIA which has been discussed in Sect. 4.4.1 of the previous chapter. As a true differential design with a single current source is not compatible with the concept of frequency dependent source degeneration to implement the high-pass filters of the equalizer[3], a differential design with two current sources has been selected. Notice that the common-mode resistance at the source of the input transistors is very high because it is determined only by the small-signal output resistance of $M_{oc,a}$ and $M_{oc,b}$. This results in a low voltage gain for common-mode signals because the source of the input transistors is heavily degenerated. In order to reject common-mode signals even more, a common-mode rejection circuit is provided. The common-mode voltage at the source of the input transistors $M_{1,a}$ and $M_{1,b}$ is measured by means of $R_{cm,a}$ and $R_{cm,b}$ and inverted by $M_{cm,1}$ and $M_{cm,2}$. The inverted common-mode voltage is applied at the gate of the current sources $M_{cm,a}$ and $M_{cm,b}$ which regulate the common-mode output voltage.

[3] The concept of frequency dependent source degeneration requires that a small-signal voltage can exist at the source of the input transistors. This is in contradiction with the concept of a differential stage where the differential small-signal voltage at the source of the input transistors is zero by definition.

The equalizer in Fig. 5.7 implements three high-pass filters which are respectively realized by $M_{R_1,a}$, $M_{R_1,b}$ and C_1, $M_{R_2,a}$, $M_{R_2,b}$ and C_2 and C_3. The effect of this tuning network is sensed only by differential-mode signals because it is placed between the sources of the input transistors and not between a source and ground. The required resistor values are consequently 2 times larger compared to a single-ended equalizer implementing the same transfer function. By contrast, the required capacitor values are 2 times smaller. In order to be able to change the equalizer frequency characteristic during operation, nMOS transistors in their linear region are used instead of resistors in the tuning network. While the first two high-pass filters are implemented by a dedicated resistance and capacitance, the third filter is implemented with the capacitor C_3 alone, as can be seen in Fig. 5.7. Instead of being determined by the product of the dedicated resistance and capacitance which is the case for the other two filter branches, the corner frequency of the third filter branch is set by C_3 and the resistance it experiences at its terminals. The latter is determined mainly by the transconductance of the input transistors. Although it was concluded in Sect. 5.1 that two filter branches are sufficient to equalize a differential n-well/p-sub photodiode up to 5 GHz, a third filter branch is used in this design to compensate for the pole at the output nodes of the equalizer which is formed by its output resistance and its total load capacitance.

The DC voltage gain for differential-mode input signals is set by the ratio of the small-signal resistance at the output nodes and the small-signal differential resistance at the source of the input transistors. While the former is determined almost completely by $M_{2,a}$ and $M_{2,b}$, which are biased in their linear region, the latter is determined predominantly by $R_{cm,a}$ and $R_{cm,b}$.

This equalizer has been designed in a 130 nm CMOS technology. The simulated Bode plot for differential-mode input signals is shown in Fig. 5.8. For frequencies below approximately 5 GHz, the magnitude response in Fig. 5.8a resembles the ideal response, as represented in Fig. 5.4a. For higher frequencies, however, the pole at the output node starts to determine the frequency characteristic which can be seen

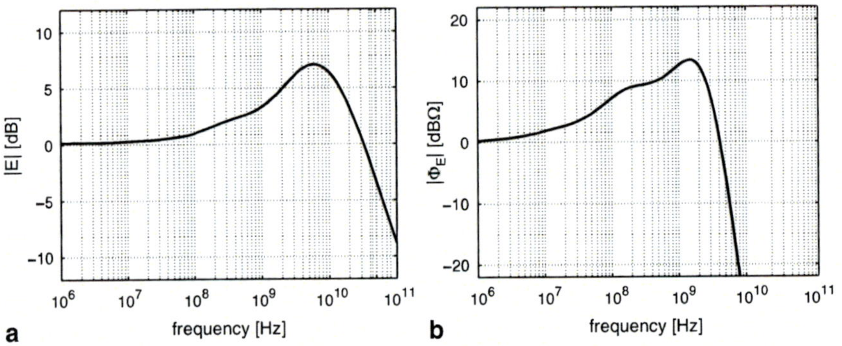

Fig. 5.8 Bode plot of the differential equalizer of Fig. 5.7 for differential-mode input signals. **a** Magnitude. **b** Phase

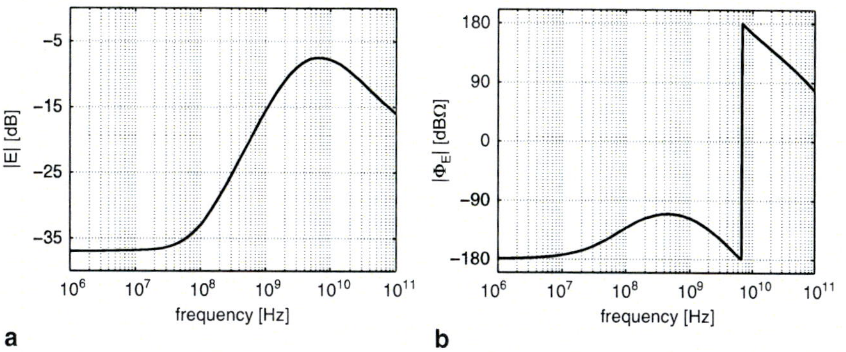

Fig. 5.9 Bode plot of the differential equalizer of Fig. 5.7 for common-mode input signals. **a** Magnitude. **b** Phase

clearly in the Bode plot as the −20 dB/decade roll-off for frequencies above 5 GHz. Common-mode signals experience a completely different transfer characteristic, as can be seen in the Bode plot in Fig. 5.9. Thanks to the large common-mode resistance at the source of the input transistors and the common-mode rejection circuit, these signals are attenuated over the entire frequency range. This means that the equalizer also contributes to the subtraction of the current of the non-illuminated photodiode from that of the illuminated photodiode. The current consumption of this equalizer is 6.39 mA, of which 5.40 mA flows through the two input transistors and 993 μA is reserved for the common-mode rejection circuit. This current consumption cannot be reduced to a lower value because the pole at the output nodes would shift to too low a frequency and thereby reduce the efficacy of the equalizer.

5.3.2 A Single-Ended Equalizer with Active Inductors

The next equalizer design which is presented here aims to extend the bandwidth of a single-ended n-well/p-sub photodiode up to a frequency of approximately 500 MHz. As will be explained in Chap. 7, this equalizer is intended to be used together with the TIA that has been discussed in Sect. 4.4.2 and is therefore designed in the same 180 nm CMOS technology. For the same reason, it is operated at the same low supply voltage of 1.12 V.

The circuit diagram of this equalizer is depicted in Fig. 5.10. Three active inductors are placed in parallel to constitute the load network of transistor M_e. By making use of *triple-well* transistors for M_1, M_2 and M_3, the *bulk effect* has been eliminated by connecting their bulk and source to each other. The voltage gain of this circuit at low frequencies is given by the following formula:

$$E_{LF} = -\frac{g_{m,M_e}}{g_{m,M_1} + g_{m,M_2} + g_{m,M_3} + g_{ds,M_e} + g_{ds,M_1} + g_{ds,M_2} + g_{ds,M_3}}, \quad (5.20)$$

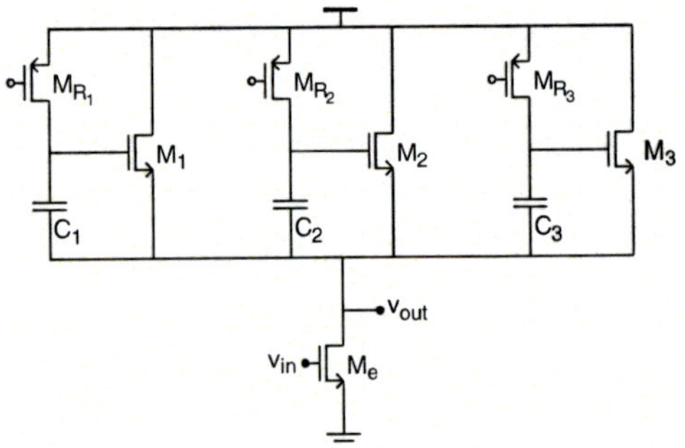

Fig. 5.10 Circuit diagram of the single-ended equalizer with active inductors in 180 nm CMOS

where g_{ds,M_i} is the reciprocal of r_{ds,M_i} and thus represents the small-signal output conductance of M_i which is biased in the saturation region. The sum of the currents through the three load transistors equals the current through the input transistor. If all nMOS transistors are biased with the same overdrive voltage, $|E_{LF}|$ is smaller than 1 due to the presence of the output conductances in (5.20). Therefore, in order to realize a gain of 1 at low frequencies, the overdrive voltage of the three load transistors should be made slightly higher than the overdrive voltage of the input transistor. This is obtained by setting the DC voltage at the input, which determines the overdrive voltage of M_e, slightly lower than the voltage difference between the supply voltage and the DC voltage at the output, which determines the overdrive voltage of M_1, M_2 and M_3. Logically, the DC voltage at the input is determined by the DC output voltage of the previous stage, namely the TIA of Sect. 4.4.2. Therefore, the DC input voltage of this equalizer equals 0.56 V, resulting in an overdrive voltage of 50 mV for M_e. The voltage gain at high frequencies is determined by the following formula:

$$E_{HF} = -\frac{g_{m,M_e}}{g_{ds,M_e} + g_{ds,M_1} + g_{ds,M_2} + g_{ds,M_3} + G_{M_{R_1}} + G_{M_{R_2}} + G_{M_{R_3}}}, \tag{5.21}$$

where $G_{M_{R_i}}$ is the conductance of transistor M_{R_i} which is biased in its linear region[4]. While the value of $G_{M_{R_i}}$ can be regulated by changing the gate voltage of the respective pMOS, this is not the case for the small-signal output conductances of the nMOS transistors which are constant. As a result, the value of (5.21) is maximized if the resistance of M_{R_i} is maximized or, equivalently, when the gate-source voltage of the pMOS transistors is minimized.

The sizes of M_1, M_2 and M_3 in this design are the same, which means that they conduct the same current, namely 88.5 µA, and consequently have the same transconductance and output impedance. As a consequence, the difference between the filter branches is determined only by the capacitances of the capacitors C_1, C_2 and C_3 and the resistances of the pMOS transistors which also have the same size. The most important design parameters of this equalizer are summarized in Table 5.2. Using (5.20), it can be concluded from this table that the magnitude of the gain at

[4] $G_{M_{R_i}}$ is the reciprocal of $R_{M_{R_i}}$, as defined in (4.99).

Table 5.2 Key design parameters of the single-ended equalizer in Fig. 5.10

		g_{m,M_e}	3.96 mS		
		g_{ds,M_e}	168 µS		
g_{m,M_1}	1.23 mS	g_{m,M_2}	1.23 mS	g_{m,M_3}	1.23 mS
g_{ds,M_1}	52.9 µS	g_{ds,M_2}	52.9 µS	g_{ds,M_3}	52.9 µS
C_1	2 pF	C_2	0.5 pF	C_3	0.125 pF
$R_{M_{R_1}}$	813 Ω–2.46 GΩ	$R_{M_{R_2}}$	813 Ω–2.46 GΩ	$R_{M_{R_3}}$	813 Ω–2.46 GΩ

low frequencies equals 0.986, which is close enough to 1. As can also be seen in this table, the resistance of the pMOS transistors can vary over a huge range by altering their gate voltages between the supply voltage and ground. The equalizer frequency characteristic for different resistance settings is shown in Fig. 5.11. Notice that the higher resistance setting boosts the magnitude response of the equalizer at lower frequencies. For the smallest resistance setting, no gain boosting can be seen in Fig. 5.11**a and the equalizer acts as an inverting amplifier with a gain of over the** entire frequency range. The nominal resistance setting boosts only the frequency components which are attenuated by the photodiode. Characteristics between the two extreme cases in Fig. 5.11 can be obtained by tuning the gate voltages of the pMOS transistors, which can be done during operation. The unit step response of this equalizer is shown in Fig. 5.12. As expected, the response for the smallest resistance setting is very close to the applied step function. For the largest resistance setting the equalizer response cannot be represented completely on the limited time scale of Fig. 5.12 because it takes several microseconds to decay to an amplitude of 1 V. The unit step response for the nominal resistance setting should be as close as possible to the inverse unit step response of the integrated photodiode.

Thanks to the low overdrive voltage of the nMOS transistors, the total power consumption of this circuit is only 297 µW.

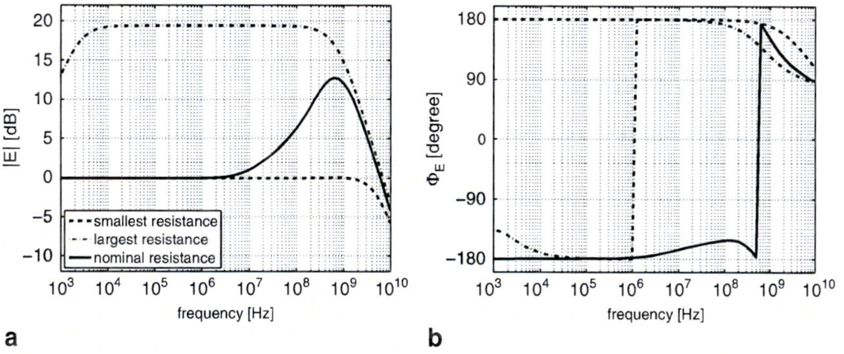

Fig. 5.11 Bode plot of the single-ended equalizer in Fig. 5.10 for different settings of the tuning network. **a** Magnitude. **b** Phase

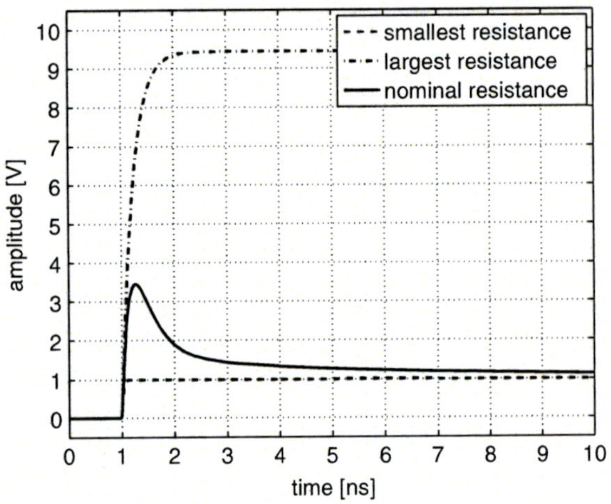

Fig. 5.12 Unit step response of the single-ended equalizer in Fig. 5.10 for different settings of the tuning network

5.4 Conclusion

As discussed in Chap. 3, a photodiode which is integrated in a CMOS process suffers from a serious bandwidth limitation regarding the conversion of light into an electrical current. The equalizer aims at extending this bandwidth in order to enable data communication at much higher bit rates than what would be possible without equalization.

The operating principle of an equalizer is examined first. It is shown that an ideal equalizer should have the inverse frequency response of the integrated photodiode. Such a response with a low magnitude roll-up can be realized by placing several high-pass filters with prescribed gains in parallel with a direct branch. In order to compensate a differential n-well/p-sub photodiode up to a frequency of 5 GHz, at least two high-pass filters are needed, as can be concluded after executing an optimization algorithm.

Two possibilities are presented thereafter that realize a high-pass filter with prescribed gain. The first circuit is a source-degenerated amplifier in which the impedance at the source of a transistor drops as the frequency rises. The second circuit uses an inductive load network which exhibits an increasing impedance as the frequency rises. It is also shown that a complete equalizer consisting of a direct branch and several high-pass filters can be realized by adapting the basic source-degenerated amplifier and inductively loaded amplifier.

Finally, two circuit designs are shown. The first design is designed in a 130 nm CMOS process and uses the principle of frequency dependent source degeneration while the second design is designed in a 180 nm CMOS process and applies an active inductive load network to realize the desired frequency response.

Chapter 6
Towards a Rail-to-Rail Voltage—The Post Amplifier

Although the photocurrent is amplified significantly by the TIA, the signal swing at the output of the equalizer is still too low to be considered as digital data. The post amplifier tackles this last analog task of an optical receiver by increasing the signal swing to a rail-to-rail level.

The difference between a limiting amplifier and an automatic gain control amplifier is explained first. The advantages and disadvantages of both are discussed and it is concluded that a limiting amplifier is the better choice for the intended application.

The specifications of a post amplifier are summarized in the following section. Most importantly, these are the small-signal gain and the related power penalty, the bandwidth and the related group delay variation, the noise, the input dynamic range and the offset voltage.

Some broadband circuit techniques are discussed thereafter. As the post amplifier needs to deliver its gain over a very wide frequency range, conventional gain boosting techniques are not applicable.

Next, it is shown how the offset voltage can be reduced by applying a feedback loop over the complete amplifier. A solution is also proposed to realize in an efficient manner the large time constant which is required in this feedback loop.

To end the chapter, two circuit designs are examined.

6.1 A Limiting Amplifier or an AGC Amplifier?

In the literature, two main classes can typically be distinguished regarding post amplifiers. The first class comprises *limiting* amplifiers which have a fixed gain. The second class comprises *automatic gain control* (AGC) amplifiers which, as the name already suggests, vary their gain automatically. This difference between the two types can best be described by their input-output characteristics which are shown schematically in Fig. 6.1. As can be seen in Fig. 6.1a, a limiting amplifier has a very large gain which cannot be altered. As a result, only an input signal with a moderate swing is amplified linearly as the output voltage cannot rise unrestrictedly due to the finite supply voltage. An input signal with a larger swing is amplified in a highly

F. Tavernier, M. Steyaert, *High-Speed Optical Receivers with Integrated Photodiode in Nanoscale CMOS,* Analog Circuits and Signal Processing 5,
DOI 10.1007/978-1-4419-9925-2_6, © Springer Science+Business Media, LLC 2011

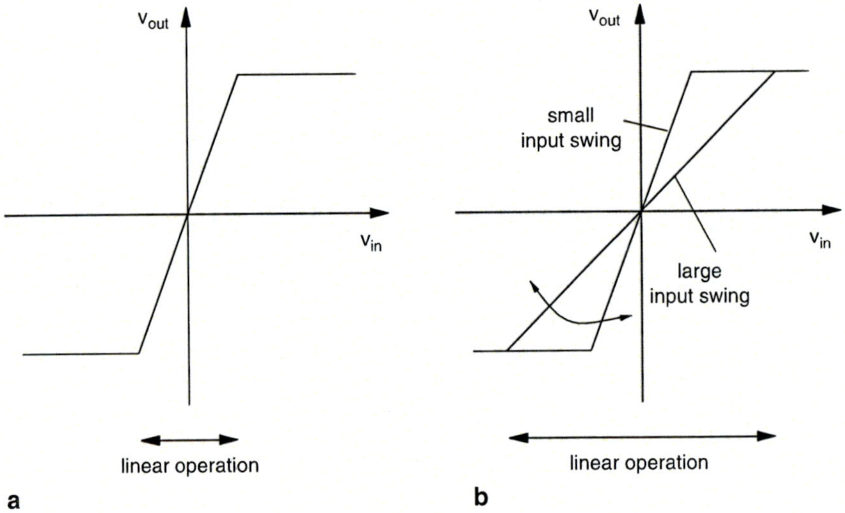

Fig. 6.1 **Input-output characteristic of a limiting amplifier and an AGC amplifier. a** A limiting amplifier has a large and fixed voltage gain. **b** An AGC amplifire modifies its voltage gain depending on the input signal swing

non-linear way which introduces significant harmonic distortion. By contrast, an AGC amplifier typically consists of a *variable gain amplifier* (VGA) with a feedback loop that controls the gain. While a limiting amplifier limits the output signal swing for an input signal that is too large, an AGC amplifier reduces its gain so that it can also provide linear operation for larger input signals. Logically, every AGC amplifier has a minimal gain setting which implies that clipping eventually occurs for the largest input signals.

From the discussion above, it seems that an AGC amplifier outperforms a limiting amplifier because it provides better linearity. However, there are a lot of other performance issues that have not been considered until now. Thanks to the lack of a gain control mechanism, a limiting amplifier requires a less complex circuit and is thereby easier to design and furthermore has a lower power consumption. Moreover, it is easier to obtain a higher small-signal bandwidth and better noise performance with a limiting amplifier [Säc05]. The performance differences between the two amplifier types are summarized in Table 6.1.

Table 6.1 Performance summary of a limiting amplifier and an AGC amplifier

	Limiting amplifier	AGC amplifier
Linearity	−	+
Complexity	+	−
Bandwidth	+	−
Noise	+	−
Power consumption	+	−

The decision as to whether a limiting amplifier or an AGC amplifier should be used depends largely on the following observations:

- No analog signal processing is required for the output signal of the post amplifier;
- The transmitted signal has a digital format with only two possible signal levels.

The consequence of these observations is that linearity is not an important issue. However, bandwidth, noise and power consumption are always important issues. A limiting amplifier has therefore been chosen in this work.

6.2 Important Specifications

Although the performance of the optical receiver depends mainly on the behavior of the first building blocks, namely the photodetector and the TIA, the post amplifier can ruin the receiver's performance. The most significant specifications of a post amplifier in an optical receiver are therefore discussed below. As a limiting amplifier is used in this work, the discussion is restricted to this particular kind of amplifier.

Gain and Power Penalty By definition, the gain of a limiting amplifier is the variation of the output voltage normalized to the variation of the input voltage. If it can be assumed that the signals at the input as well as at the output are small enough so that the limiting amplifier stays in its linear region, this definition is equivalent to the following formula:

$$A = \frac{v_{out}}{v_{in}}, \tag{6.1}$$

where v_{in} and v_{out} represent the small-signal input and output voltage, respectively. Typically, a limiting amplifier has a differential implementation with differential inputs and outputs, as shown in Fig. 6.2. The input voltage in (6.1) is then given by $v_{in} = v_{in,p} - v_{in,n}$ while the output voltage is given by $v_{out} = v_{out,p} - v_{out,n}$. The gain is then called the *differential* gain. If only one of the outputs is used, the gain is called the *single ended* gain and is a factor of 2 lower than the differential gain.

The gain as defined in (6.1) is usually specified only for small input signals where the limiting amplifier is operating in its linear region. The value of A is then actually the same as the slope of the curve in Fig. 6.1a for an input voltage around 0 V. For larger input signals, the gain is compressed due to the limiting behavior. It can be understood that the equivalent value of A is lower for such signals. If the input signal has a swing which is already rail-to-rail, the voltage gain of the limiting amplifier actually decreases to 1.

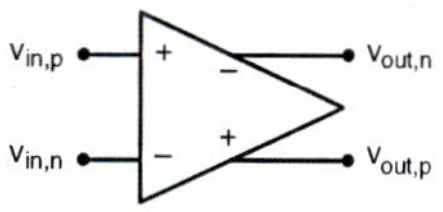

Fig. 6.2 Inputs and outputs of a differential limiting amplifier

Ideally, the small-signal gain of a limiting amplifier should be infinite. Such an infinite gain is not realizable in practice, which means that there is a certain linear region of operation in which signals are not amplified completely to a rail-to-rail voltage. The limiting amplifier is followed typically by a *decision circuit* which decides whether a ZERO or a ONE has been transmitted. If the decision circuit is perfect, the finite gain of the limiting amplifier does not introduce errors. However, a real decision circuit is always subjected to some noise which introduces errors if its input voltage swing is below a certain sensitivity level V_s. This non-ideality can be expressed mathematically in terms of a power penalty PP as follows [Säc05]:

$$PP = 1 + \frac{V_s}{v_{out}}, \tag{6.2}$$

where v_{out} is the peak-to-peak value of the signal at the output of the limiting amplifier. Logically, the power penalty is maximal if v_{out} has its minimal swing, namely when v_{in} is minimal. This minimal limiting amplifier input swing depends on the noise of the receiver and the gain of the TIA:

$$v_{in,min} = 2 \cdot 7 \cdot Z_{TIA,DC} \cdot i_{n,in,RMS}, \tag{6.3}$$

where the factor 7 is determined by the Q-function for a BER of 10^{-12} (Sect. 2.5.3) and $Z_{TIA,DC}$ and $i_{n,in,RMS}$ are the DC transimpedance and the input-referred RMS noise current of the TIA, respectively. It is assumed in (6.3) that the noise of the receiver is only determined by the noise of the TIA, which is a reasonable assumption if the transimpedance is large enough. The maximal power penalty due to the finite sensitivity of the decision circuit can then be expressed as follows:

$$PP = 1 + \frac{V_s}{2 \cdot 7 \cdot A \cdot Z_{TIA,DC} \cdot i_{n,in,RMS}}, \tag{6.4}$$

where A represents the small-signal voltage gain of the limiting amplifier. This expression can be solved for A:

$$A \geq \frac{V_s}{(PP - 1) \cdot 2 \cdot 7 \cdot Z_{TIA,DC} \cdot i_{n,in,RMS}}. \tag{6.5}$$

It can be concluded from this formula that the limiting amplifier should actually provide no gain at all if a perfect decision circuit is available ($V_s = 0$ V). However, for a real decision circuit, the gain should have a certain minimal value which is relaxed by having a TIA with a large transimpedance. From (6.5), it could also be concluded that a larger input-referred noise in the TIA relaxes the limiting amplifier gain. However, this larger noise would jeopardize the receiver sensitivity directly.

If the sensitivity of the decision circuit is 10 mV, the transimpedance equals 1 kΩ, the input-referred noise of the TIA is 1 μA_{RMS} and a power penalty of 0.5 dB = 1.12 can be tolerated, the minimal gain of the limiting amplifier can be calculated with (6.5) which results in 15.3 dB = 5.85. On the contrary, if a power penalty of only 0.05 dB = 1.01 can be tolerated, the required gain of the limiting amplifier increases to 35.5 dB = 59.5.

Bandwidth and Group Delay Variation The small-signal voltage gain A cannot be maintained over an infinite frequency range. The bandwidth is defined as the frequency where the gain is 3 dB lower than at low frequencies. Like the voltage gain, this figure is a small-signal quantity. This is due to the non-linear behavior of transistors for signals with a large swing which causes their output resistance to alter drastically over the complete signal range. As concluded in Sect. 2.5.3, the bandwidth of a system should be 0.6–0.7 · B Hz if the bit rate equals B bit/s. As a limiting amplifier is typically operated in its limiting regime, this small-signal bandwidth is only of marginal significance. Another measure should therefore be used to quantify the speed performance of a limiting amplifier when input signals are applied with a relatively large swing. However, it is generally known that the small-signal bandwidth is a rather conservative measure for the speed performance which is due to the switching behavior of a circuit when a large signal is applied [Säc05]. This can be seen in Fig. 6.3, where the result of a transient simulation of a double common-source amplifier is shown for a square wave input signal with a small and a large swing. The simulated −3 dB bandwidth of this double common-source amplifier is 645 MHz which is a perfectly acceptable value for the 1 Gbit/s input signal. The settling of the output voltage is then more-or-less completed in a single bit interval if the input signal is small (Fig. 6.3a). By contrast, the settling time is significantly reduced if the input signal has a much larger swing (Fig. 6.3b). This phenomenon can also be quantified by considering the time it takes for the output voltage to reach the threshold voltage of 0.6 V, half the supply voltage. If the input signal is small, this delay can be calculated as follows:

$$t_{small} = -\frac{\ln 0.5}{2\pi BW} = 0.693 \cdot R_{out} C_{out}, \tag{6.6}$$

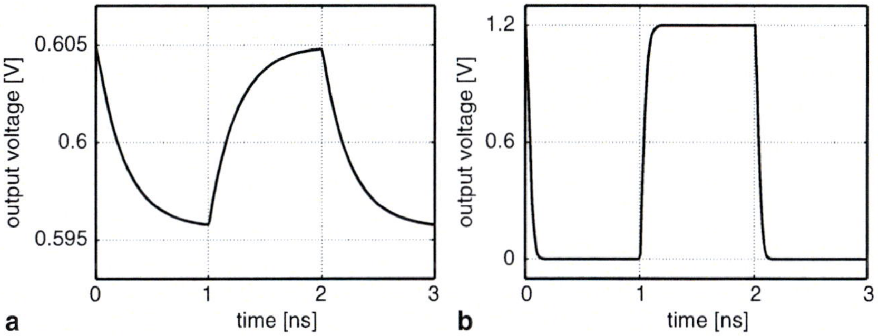

Fig. 6.3 Transient simulation results for a double common-source amplifier in a 130 nm CMOS process where both transistors have a minimal length of 120 nm and where the nMOS and pMOS have widths of 1 μm and 2.69 μm, respectively. The double common-source amplifier is connected to a load capacitance of 30 fF and the supply voltage is 1.2 V. **a** Evolution of the output voltage when the input signal is a square wave with an amplitude of 1 mV and a bit rate of 1 Gbit/s. **b** Evolution of the output voltage when the input signal is a square wave with an amplitude of 1.2 V and a bit rate of 1 Gbit/s

where R_{out} is the small-signal output resistance of the double common-source amplifier and C_{out} is its total output capacitance. As already mentioned, the bandwidth of the simulated double common-source amplifier in Fig. 6.3 equals 645 MHz, which results in a calculated delay of 171 ps between the start of the bit interval and the time when the output voltage reaches half of its final value. By contrast, for a large input signal, the following formula can be used to estimate the same delay [Rab03]:

$$t_{large} = 0.693 \cdot \left(\frac{R_{eq,n} + R_{eq,p}}{2}\right) C_{out}, \tag{6.7}$$

where $R_{eq,n}$ and $R_{eq,n}$ represent the average output resistance of the nMOS and pMOS, respectively, if the output voltage evolves towards the threshold voltage from its highest and lowest values, respectively. To take into account the difference in the rise and fall time of the double common-source amplifier, the resistances of the nMOS and pMOS are averaged, as can be seen in (6.7). If the input voltage swing is rail-to-rail, these resistances can be calculated as follows [Rab03]:

$$R_{eq,n} \approx \frac{3}{4} \frac{V_{DD}}{I_{ds,sat,n}} \left(1 - \frac{5}{6} \lambda_n V_{DD}\right), \tag{6.8}$$

$$R_{eq,p} \approx -\frac{3}{4} \frac{V_{DD}}{I_{ds,sat,p}} \left(1 - \frac{5}{6} \lambda_p V_{DD}\right), \tag{6.9}$$

where $I_{ds,sat,n}$ and $I_{ds,sat,p}$ are the saturation currents of the nMOS and pMOS, respectively, for a gate-source input voltage which is equal to the supply voltage V_{DD}. The constants λ_n and λ_p represent the channel-length modulation coefficients of the nMOS and pMOS. Without giving the exact values of the equivalent resistances, the simulated average time delay of the double common-source amplifier for large input signals equals only 42.5 ps, which is more than 4 times less than for small input signals. Comparing the expressions for the time delays given by (6.6) and (6.7), it can be concluded that the longer delay for small input signals is due to the larger output resistance in the small-signal regime, because the output capacitance is the same in both expressions. Consequently, it can be concluded that the small-signal bandwidth is a conservative measure of the speed performance of a limiting amplifier.

Related to the bandwidth is the *group delay* which is the transit time of a signal through a device or circuit versus frequency. Ideally, this measure should have a flat characteristic so that every signal component is delayed equally. A mathematical expression for the group delay t_{gd} is as follows:

$$t_{gd}(\omega) = -\frac{d\Phi}{d\omega}, \tag{6.10}$$

where Φ is expressed in radians and represents the frequency dependent phase of the device or circuit under consideration and ω is the angular frequency ($\omega = 2\pi f$). If a circuit does not contain any poles or zeros, so that its phase does not change with frequency, the group delay is constant and equal to 0 s. A circuit with a single time

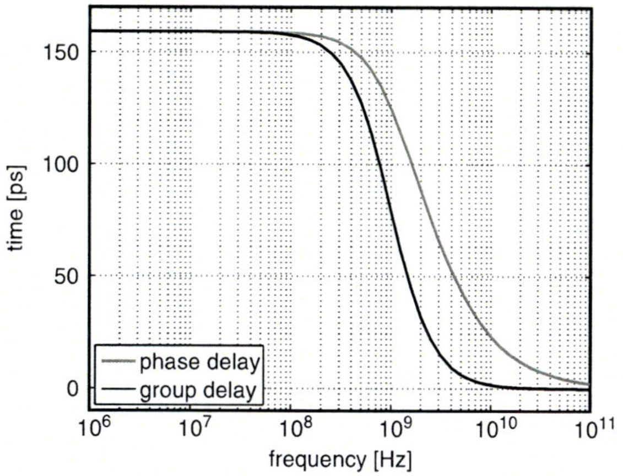

Fig. 6.4 The phase and group delay of a first-order low-pass filter with a corner frequency at 1 GHz

constant τ generating a single pole has a frequency dependent group delay which can be expressed by the following formula:

$$t_{gd}(\omega) = \frac{\tau}{1 + \tau^2\omega^2}. \tag{6.11}$$

The group delay should not be confused with the phase delay which was discussed in Sect. 4.1. While the phase delay is the delay of a single sinusoid, the group delay is a measure for the delay of a signal which is composed of multiple sinusoids. The difference between the two concepts is shown in Fig. 6.4 for a low-pass filter with a corner frequency at 1 GHz. It can be noticed that the phase and group delay are equal to each other if the frequency is considerably lower than the corner frequency where the phase varies linearly with frequency. The delay at these low frequencies is equal to the time constant of the first-order filter, 159 ps in this example. At very high frequencies, both the phase and group delay evolve towards 0 ps.

More important than the actual value of the group delay is its variation with frequency because this introduces DDJ. A typical guideline is to keep the group delay variation of a limiting amplifier below 10% of the bit interval over the entire bandwidth [Säc05]. The bandwidth of 1 GHz in the example above has been considered in Chap. 4 to be sufficient to receive data with a bit rate of 1.6 Gbit/s and a bit period of 625 ps. As can be seen in Fig. 6.4, the group delay at 1 GHz has shifted by 79.6 ps from the group delay at low frequencies, which is more than the tolerable 10% of the bit interval. For this reason, it is common to design a limiting amplifier with a bandwidth of 1–1.2 · B Hz to amplify a signal with a bit rate of B bit/s [Säc05]. Although it has been concluded in the previous paragraph that the bandwidth is a conservative measure for the operating speed of a limiting amplifier when it is used in the limiting regime, it is too optimistic a measure when considering the DDJ it introduces.

Noise Figure The noise of the limiting amplifier adds to the total noise of the optical receiver. It has been assumed until now that the receiver noise is determined completely by the TIA noise; this assumption is valid only if the transimpedance of the TIA is large enough and the noise of the limiting amplifier is not excessively large. This can be seen in the following formula:

$$i_{n,RMS} \approx \sqrt{i^2_{n,RMS,TIA} + \frac{v^2_{n,RMS,LA}}{Z^2_{TIA,DC}}}, \tag{6.12}$$

where $i_{n,RMS}$ is the input noise current of the complete optical receiver, $i_{n,RMS,TIA}$ is the input noise current of the TIA, $v_{n,RMS,LA}$ is the input noise voltage of the limiting amplifier and $Z_{TIA,DC}$ is the DC transimpedance of the TIA. Although the required bandwidth of the limiting amplifier is large, which leads to a relatively large integrated noise, the transimpedance of the TIA is typically large enough so that the noise of the limiting amplifier does not jeopardize the sensitivity of the optical receiver.

Input Dynamic Range The input dynamic range is the ratio of the largest to the smallest input signals for which the BER stays below a certain threshold, typically 10^{-12}. As already mentioned in the beginning of this section, the smallest input signal is defined by the gain of the limiting amplifier and the sensitivity of the decision circuit. By contrast, the largest signal is on the one hand limited by the supply voltage and on the other hand determined by the maximum tolerable pulse-width distortion.

Offset Voltage Ideally, the output signal of a differential amplifier has an output voltage of 0 V if no signal is applied at its input. In reality, however, this is not the case, as can be seen in Fig. 6.5a. An offset voltage $v_{os,out}$ appears at the output when the input is shorted. Such an offset voltage has two important effects in a limiting amplifier: a shift of the threshold voltage and the introduction of pulse-width distortion. The input offset voltage $v_{os,in}$ shown in Fig. 6.5b is, by definition, the differential input voltage which is required to make the differential output voltage zero. For an amplifier working down to DC it can be calculated as follows:

$$|v_{os,in}| = \frac{|v_{os,out}|}{A}, \tag{6.13}$$

where A denotes the small-signal gain of the differential amplifier. Offset in a differential amplifier originates from mismatch between two branches of a differential structure which are intended to be equal. Mismatch is caused by inaccuracies during

Fig. 6.5 Offset voltage of a differential amplifier. **a** The output offset voltage when no input signal is applied. **b** The input offset voltage reduces the output offset voltage to 0 V

the manufacturing process of an IC that introduce discrepancies between the wanted and realized lengths and widths of the transistors, variations in the threshold voltage of different transistors on the same chip, dissimilarities between the resistances of identical resistors, etc. [Pel89]. In [San06], it is deduced that the input offset voltage of a differential amplifier with load resistors R_l can be calculated as follows:

$$v_{os,in} = \frac{V_{gs} - V_{th}}{2} \left(\frac{\Delta R_l}{R_l} + \frac{\Delta K'}{K'} + \frac{\Delta (W/L)}{(W/L)} \right) + \Delta V_{th}, \tag{6.14}$$

where V_{gs}, V_{th}, W and L are the gate-source voltage, threshold voltage, width and length, respectively, of the input transistors. The parameter K' depends on the mobility, C_{ox}, and the subthreshold gate coupling coefficient [San06]. From this formula, it can be concluded that the offset voltage decreases with a decreasing overdrive voltage. From an offset point of view, it is therefore best to design the input transistors of a differential amplifier with a low overdrive voltage.

As discussed in detail in Sect. 2.5.3, the BER of a receiver depends strongly on the difference between the high level of the signal V_h and the threshold voltage V_t on the one hand and the low level of the signal V_l and the same threshold voltage on the other hand. If the threshold voltage is located exactly in the middle between V_h and V_l, the BER can be calculated using (2.37). However, this formula is no longer valid if the threshold voltage is shifted with respect to its ideal value, which is exactly what happens when an offset voltage is present at the output of the limiting amplifier. If $v_{os,out}$ is positive, the threshold voltage increases by $v_{os,out}$ so that it is located closer to the high level of the signal than to the low level. For the same signal swing, this leads to an increased error probability for a ONE but a decreased error probability for a ZERO. The total BER is still determined by the following formula if a ZERO and a ONE occur with the same probability:

$$P_e = \frac{1}{2} P_{e,\text{ZERO}} + \frac{1}{2} P_{e,\text{ONE}}. \tag{6.15}$$

The PDF of the voltage at the output of the limiting amplifier with an output offset voltage $v_{os,out}$ is shown in Fig. 6.6. Without giving the mathematical derivation to calculate the value of $P_{e,\text{ZERO}}$ and $P_{e,\text{ONE}}$, it can be seen in this figure that the error probability when receiving a ZERO decreases to a lesser degree than the error probability when receiving a ONE rises. Compared to Fig. 2.18b, where the threshold voltage has its optimal value, the shaded area in Fig. 6.6 representing the BER is obviously larger. The same conclusion can be drawn if the offset voltage is negative and the threshold voltage is shifted towards V_l instead of V_h. In order to maintain the same BER in the presence of an offset voltage, it is stated in [Säc05] that the signal swing at the output of the limiting amplifier should be increased by $2v_{os,out}$.

The second effect of an offset voltage pulse-width distortion of the limiting amplifier output signal which is caused by finite rise and fall times. The pulse duration for a ONE decreases if the offset voltage is positive whereas it increases if the offset voltage is negative. However, as the limiting amplifier is typically operated in its limiting regime in which the rise and fall times of the output voltage are very short, the pulse duration of a ZERO and a ONE are only marginally affected.

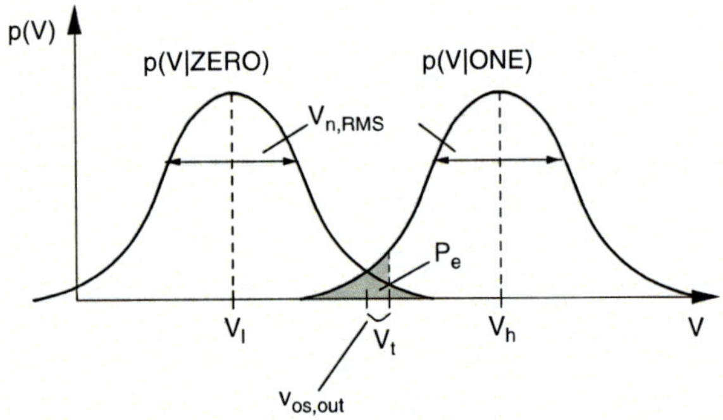

Fig. 6.6 PDF of the voltage at the output of the limiting amplifier in the presence of noise and a positive offset voltage

In [Säc05], it is calculated that the input offset voltage of the limiting amplifier in an optical receiver should be less than 1 mV in order not to jeopardize the BER performance. In order to obtain such a small value, some form of *offset compensation* should be applied. Several circuit techniques which realize this are discussed in Sect. 6.4.

6.3 Broadband Circuit Techniques

It has been concluded in the previous section that the small-signal bandwidth of the limiting amplifier in an optical receiver should be 1–1.2 · B Hz if the bit rate equals B bit/s. This bandwidth is approximately 2 times larger than the bandwidth which is needed for the TIA operating at the same bit rate. Thus it seems that the speed limitation of an optical receiver is caused mainly by the limiting amplifier as this building block needs to have the highest bandwidth. However, as discussed in Chap. 4, the bandwidth limitation of the TIA is caused mainly by the large photodiode capacitance at its input, especially if an integrated photodiode is used. The 2 times larger bandwidth of the limiting amplifier is consequently not as difficult to achieve compared with the difficulty of realizing a high TIA bandwidth.

6.3.1 Multistage Amplifier

Typically, the gain of a limiting amplifier should be larger than 30 dB to minimize the associated power penalty discussed before. By contrast, the bandwidth should be around 6 GHz when used in cooperation with the TIA which has been presented in Sect. 4.4.1. This differential TIA has a simulated bandwidth of 2.85 GHz enabling bit rates up to 4.75 Gbit/s. A limiting amplifier bandwidth of 6 GHz easily satisfies the condition that has been proposed in the previous section, namely that it should be 1–1.2 · B Hz for a bit rate of B bit/s. The required *GBW* of the limiting amplifier is consequently larger than 190 GHz. If the limiting amplifier is integrated in the 130 nm CMOS technology the parameters of which are summarized in Fig. 4.20, this value is larger than the f_t of the technology[1]. A limiting amplifier with such a large GBW can therefore only be realized by using a multistage architecture.

As already calculated in Sect. 4.3.1 a cascade of n 1-stage amplifiers can have a GBW which is much larger than that of a single 1-stage amplifier. If all the stages are identical, n poles coincide at the same frequency; this can be tolerated from a stability point of view because a limiting amplifier is normally used without feedback across the stages so that it is not necessary to have a dominant pole. In Sect. 4.3.1, it has been assumed that every 1-stage amplifier has a first-order response, in which case the following relation could be derived:

$$\frac{GBW_{cascade}}{GBW_{1\text{-}stage}} = \left|A_{1\text{-}stage,DC}\right|^{n-1} \sqrt{\sqrt[n]{2}-1}, \tag{6.16}$$

which has been visualized in Fig. 4.25 for different values of $A_{1\text{-}stage,DC}$, the DC gain of a single amplifier stage. When designing a limiting amplifier which has to realize a certain DC gain and bandwidth, the problem with this approach is that the total gain of the cascade depends on the number of stages. The question that needs to be solved to design a multistage limiting amplifier with a DC gain A_{LA} and a bandwidth BW_{LA} is quite different: what is the optimum number of stages which require the least GBW per stage? In [Jin87], it is derived that the required 1-stage GBW can be calculated as follows:

$$GBW_{1\text{-}stage} = \frac{BW_{LA}}{\sqrt{\sqrt[n]{2}-1}} \sqrt[n]{A_{LA}}. \tag{6.17}$$

Normalizing this 1-stage GBW with respect to the GBW of the complete limiting amplifier results in the following formula:

$$\frac{GBW_{1\text{-}stage}}{GBW_{LA}} = \frac{A_{LA}^{\frac{1}{n}-1}}{\sqrt{\sqrt[n]{2}-1}}. \tag{6.18}$$

This relationship is shown in Fig. 6.7 for three gain requirements of the limiting amplifier. It can be noticed that the required GBW of a single stage can be reduced considerably by cascading several stages and that the effect is more pronounced the larger the required total gain. It can also be seen that there exists a fairly weak optimum where the required 1-stage GBW is minimal. This can be calculated by differentiating (6.18) with respect to n:

$$n_{opt} = \ln(2)\left[\ln\left(\frac{2\ln(A_{LA})}{\ln\left(A_{LA}^2\right)-\ln(2)}\right)\right]^{-1}. \tag{6.19}$$

For the three limiting amplifier gain requirements of Fig. 6.7, this formula reveals that n_{opt} equals 6 or 7 if $A_{LA} = 30$ dB, 7 or 8 if $A_{LA} = 35$ dB and 8 or 9 if $A_{LA} = 40$ dB.

[1] The simulated f_t of an nMOS in that 130 nm CMOS technology is 80 GHz for an overdrive voltage of 100 mV and 119 GHz for an overdrive voltage of 200 mV.

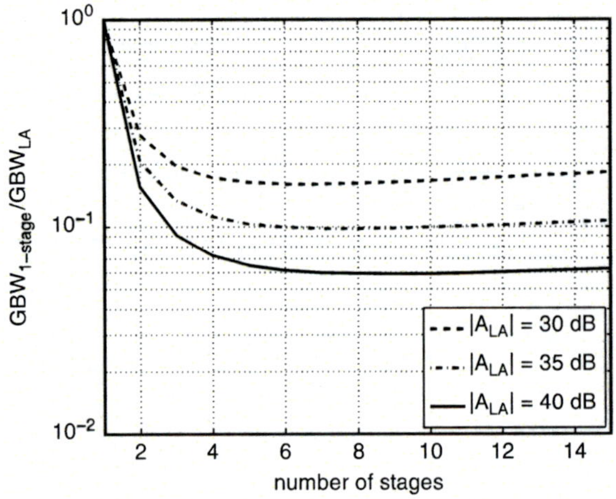

Fig. 6.7 Normalized GBW of the 1-stage amplifier for different gain requirements of the limiting amplifier

Due to the flat behavior of (6.18) around its minimum it can however also be concluded that at least four, five or six stages are needed for the three gain requirements respectively. While it was concluded previously that a GBW of more than 190 GHz in a single stage is impossible to achieve in a 130 nm CMOS technology, the required 1-stage GBW decreases to values which are considerably smaller than the f_t of the technology if a multistage limiting amplifier is used. If the limiting amplifier needs to realize a gain of 30 dB over a bandwidth of 6 GHz and if four stages are cascaded, $GBW_{1\text{-}stage}$ can be calculated with (6.18), resulting in a value of 32.7 GHz, which is indeed below the f_t of a contemporary CMOS technology.

A common-source amplifier with an input transistor M_1 which is connected to an identical common-source amplifier has the following GBW:

$$GBW_{1\text{-}stage} = \frac{g_{m,M_1}}{2\pi\left((1+\alpha)C_{gs,M_1} + C_{db,M_1}\right)}, \tag{6.20}$$

where α is a measure for the Miller effect on C_{gd,M_1} and C_{db,M_1} is the drain-bulk capacitance of M_1. If $\alpha = 1$ and assuming that $C_{db,M_1} = 0.3 \cdot C_{gs,M_1}$, the GBW of this single common-source amplifier is approximately 40% of the f_t of M_1, regardless of its sizing. Consequently, choosing the number of stages by means of (6.19) is not always appropriate because there is no advantage in reducing the required $GBW_{1\text{-}stage}$ below 40% of the transistor's f_t when a cascade of common-source amplifiers is used. Note however that f_t can be increased or decreased by increasing of decreasing the overdrive voltage.

While the design of a limiting amplifier is facilitated considerably by cascading stages, this approach has also two important drawbacks, namely an increased power consumption and a worse noise performance. The former can be understood as follows for a limiting amplifier that is loaded with a capacitance C_l and composed of cascaded common-source amplifiers. The GBW of the last stage in the cascade is determined by the following formula:

$$GBW_{1\text{-}stage} = \frac{g_{m,M_1}}{2\pi C_l}, \tag{6.21}$$

where it is assumed that C_{db,M_1} can be neglected compared to C_l. The transconductance and thereby the current and power consumption of this last stage vary linearly with $GBW_{1\text{-}stage}$ because C_l is constant. The power consumption of the complete limiting amplifier with n identical stages that each have the same $GBW_{1\text{-}stage}$ is then a function of the number of stages as can be seen in the following formula:

$$\frac{P_{LA,n}}{P_{LA,1}} = n\frac{A_{LA}^{\frac{1}{n}-1}}{\sqrt{\sqrt[n]{2}-1}}, \tag{6.22}$$

where $P_{LA,1}$ and $P_{LA,n}$ represent the power consumption of the limiting amplifier when it is realized with one stage and n stages, respectively. This relationship is shown in Fig. 6.8 for three values of A_{LA}. It can be seen that the minimal power consumption is obtained at a smaller number of stages than what is needed to minimize the GBW of a single stage as predicted by (6.19). The power consumption even rises above $P_{LA,1}$ if too many stages are cascaded.

The other drawback with a multistage amplifier is its noise performance which is always worse than that of a single stage limiting amplifier. As the number of cascaded stages is increased, the gain per stage is reduced. As the gain per stage becomes less, the noise generated by subsequent stages becomes important at some point. Taking into account the power consumption and the noise performance, it can be concluded at this point that it is best to make the number of stages slightly smaller than what results from (6.19) which is no problem because the normalized required $GBW_{1\text{-}stage}$ has a fairly flat behavior around its optimum.

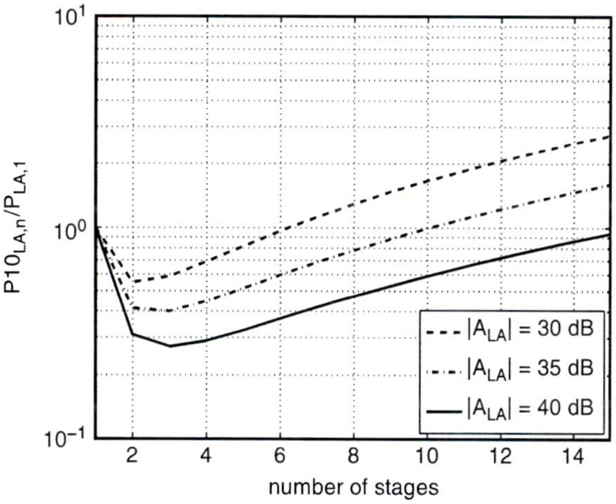

Fig. 6.8 Normalized power consumption of an n-stage limiting amplifier for different gain requirements

6.3.2 Negative Impedance Converter

The presence of a non-zero output resistance to ground and a non-zero output capacitance to the same ground causes a pole at the output of an amplifier. While the output resistance partly determines the gain of the stage as well as its bandwidth, the output capacitance only limits the bandwidth. An amplifier with better performance can consequently be realized if the output resistance is increased while the output capacitance is decreased. This is exactly the goal of using a *negative impedance converter* (NIC) [Gal03], which typically has a double functionality:

1. increase the gain by placing a negative resistance in parallel with the output resistance;
2. increase the bandwidth by placing a negative capacitance in parallel with the output capacitance.

The circuit diagram of a first-order amplifier with a NIC at its output is shown in Fig. 6.9. The output resistance R_{out} of this circuit is determined by the parallel combination of $R_{out,A}$, the output resistance of the amplifier itself, and $-R_{NIC}$, the resistance to ground of the NIC:

$$R_{out} = \frac{R_{NIC} R_{out,A}}{R_{NIC} - R_{out,A}}. \tag{6.23}$$

It can be concluded from this formula that the output resistance becomes infinitely large if $R_{NIC} = R_{out,A}$. The voltage gain from v_{in} to v_{out} is then infinitely large also, as it is determined by the transconductance g_m and the output resistance. If $R_{NIC} < R_{out,A}$ the output resistance of the circuit in Fig. 6.9 is negative, resulting in a pole which is located in the right half-plane so that the amplifier behaves as an oscillator. Therefore, in order to obtain a stable circuit with a pole in the left half-plane, R_{NIC} should always be larger than $R_{out,A}$. In addition to an increased voltage gain this larger output resistance leads to a decreased bandwidth, just like in

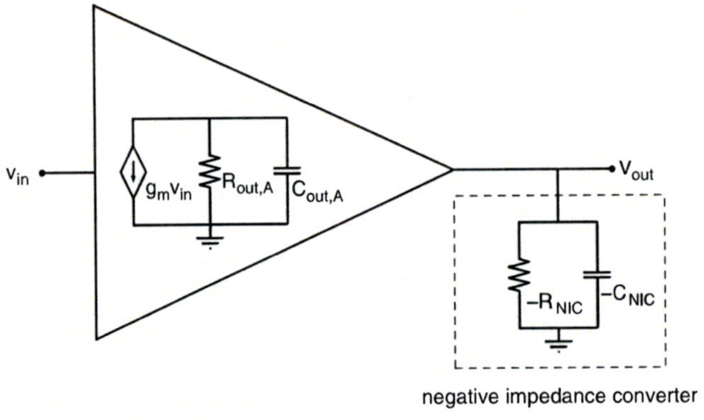

Fig. 6.9 A first-order amplifier with a NIC placed at its output

a cascode. To compensate for this, a NIC also has a capacitive part which is negative. The output capacitance C_{out} of the circuit in Fig. 6.9 can be calculated by means of the following formula:

$$C_{out} = C_{out,A} - C_{NIC}. \tag{6.24}$$

The total output capacitance becomes zero if $C_{NIC} = C_{out,A}$, in which case the bandwidth increases to infinity. A negative output capacitance is realized if $C_{NIC} > C_{out,A}$ in which case the output pole moves to the right half-plane and the amplifier behaves as an oscillator. A stable circuit with a reduced output capacitance is obtained if $C_{NIC} < C_{out,A}$.

The negative resistance and capacitance of a NIC can be realized by the circuit shown in Fig. 6.10, where it is used to compensate for the output impedance of a differential amplifier [Yoo07]. Without the addition of the NIC, the output resistance of the differential amplifier is simply determined by R_l, the resistance of $R_{l,a}$ and $R_{l,b}$. The negative resistance of the NIC is implemented by means of the cross-coupled differential pair which is composed of $M_{2,a}$ and $M_{2,b}$, each of which has a transconductance $g_{m,2}$. The differential impedance which is realized by this part of the NIC can be calculated as follows:

$$Z_{NIC,nr} = -\frac{2}{g_{m,2}}, \tag{6.25}$$

which is purely resistive. The equivalent single-ended resistance $-R_{NIC}$ of this NIC is consequently only half of this differential value. As this resistance is in parallel with the two load resistors of the differential amplifier, the resulting output resistance is as follows:

$$R_{out} = \frac{R_l}{1 - g_{m,2}R_l}. \tag{6.26}$$

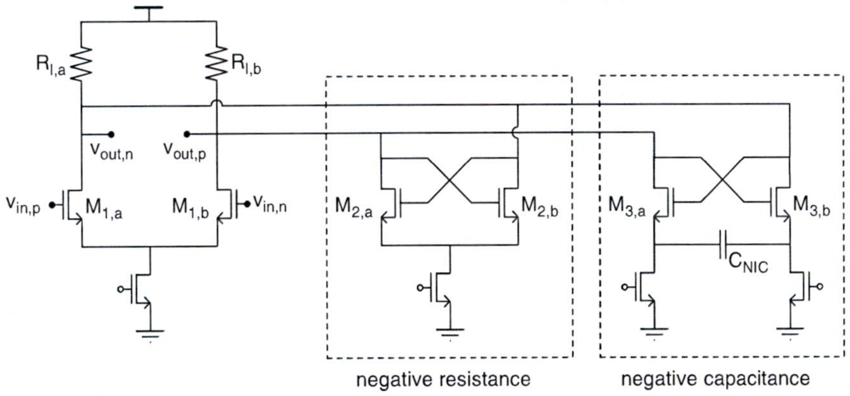

Fig. 6.10 Circuit diagram of a differential amplifier with a NIC at its output which is composed of a cross-coupled differential pair with and without frequency dependent source degeneration to implement the negative resistance and capacitance respectively [Yoo07]

In order to remain stable, $g_{m,2}R_l$ should be kept below 1 because otherwise R_{out} is negative. The negative capacitance is implemented with the cross-coupled differential pair $M_{3,a}$-$M_{3,b}$ of which the sources are degenerated by means of the current sources and capacitor C_{NIC}. The differential impedance $Z_{NIC,nc}$ that is realized by this circuit is as follows:

$$Z_{NIC,nc} = -\frac{1}{j2\pi f C_{NIC}}\left(1 + \frac{j2\pi f 2C_{NIC}}{g_{m,3}}\right). \tag{6.27}$$

In contrast to $Z_{NIC,nr}$ which is completely resistive, $Z_{NIC,nc}$ is capacitive only for frequencies below the pole frequency which is determined by $g_{m,3}$ and C_{NIC}. Its capacitance is then equal to $-C_{NIC}$, as can be seen in (6.27). For frequencies above this pole, $Z_{NIC,nc}$ is resistive and its value can be calculated with (6.25) in which $g_{m,2}$ is replaced by $g_{m,3}$. As the equivalent single-ended capacitance below the pole frequency is $-2C_{NIC}$, the resulting output capacitance can be calculated with the following formula:

$$C_{out} = C_l - 2C_{NIC}, \tag{6.28}$$

where C_l represents the single-ended load capacitance at the output of the differential amplifier. In order to obtain stable operation with a pole in the left half-plane, C_{NIC} should be smaller than half of the load capacitance so that the resulting output capacitance is positive.

Until now, it has been assumed that a NIC compensates the output resistance to increase the gain and the output capacitance to restore or increase the bandwidth. If one is only interested in increasing the bandwidth, one can choose to compensate only the output capacitance. This can be done by implementing only the negative capacitance of the NIC which is shown in Fig. 6.10. It is also possible to make use of the *negative Miller effect* to realize a negative capacitance. This technique is applied in the amplifier which is depicted in Fig. 6.11. Capacitors C_{NIC} are connected between the input and output of a differential amplifier with amplification B. In contrast to the normal Miller effect in which an inverting amplification is applied between the terminals of a capacitor, the amplification over the capacitors in Fig. 6.11 is positive. The resulting negative capacitance appears in parallel with $C_{l,A}$, the total load capacitance on the output node of the first amplifier. The output capacitance on this node can now be calculated as follows:

$$C_{out,A} = C_{l,A} + (1 - |B|)\, C_{NIC}. \tag{6.29}$$

It can be concluded from this that the output capacitance can be reduced significantly if $|B| > 1$. If $|B| = 1$, however, the addition of C_{NIC} has no effect because there is no voltage drop across these feedback capacitors. If $|B|$ approaches 0, the placement of C_{NIC} has the same effect as placing an extra load capacitor at the output which should be avoided. The following condition should be fulfilled in order to obtain stable operation:

$$C_{NIC} < \frac{C_{l,A}}{1 - |B|}. \tag{6.30}$$

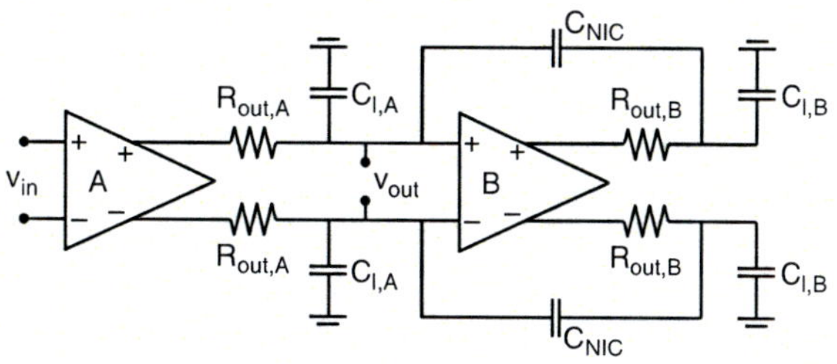

Fig. 6.11 Differential amplifier of which the load capacitance is compensated by means of negative Miller capacitors

Next to a reduction of the capacitance on the output node of the first amplifier, the addition of the feedback capacitors leads to an increase of the output capacitance of the second amplifier:

$$C_{out,B} = C_{l,B} + \left(1 - \frac{1}{|B|}\right) C_{NIC} \approx C_{l,B} + C_{NIC}. \tag{6.31}$$

Together with the output resistance $R_{out,B}$ of the second stage, this increased output capacitance generates a pole which has been neglected until now. If this pole is taken into account, the value of $C_{out,A}$ should be calculated with the following formula:

$$C_{out,A} = C_{l,A} + \left(1 - \frac{|B|}{1 + j2\pi f R_{out,B} C_{out,B}}\right) C_{NIC}. \tag{6.32}$$

At low frequencies, the capacitance at the output of the first stage is indeed reduced effectively. However, the loop gain around C_{NIC} disappears at high frequencies and $C_{out,A}$ is simply the sum of $C_{l,A}$ and C_{NIC}. In order to increase the bandwidth of the voltage gain from v_{in} to v_{out} it can be stated that the pole at the output of the second amplifier should be at a frequency which is significantly higher than that bandwidth:

$$\frac{1}{2\pi R_{out,B} C_{out,B}} \gg \frac{1}{2\pi R_{out,A} C_{out,A}} \tag{6.33}$$

$$\gg \frac{1}{2\pi R_{out,A} \left(C_{l,A} + (1 - |B|) C_{NIC}\right)}. \tag{6.34}$$

Although it has been assumed in this section that a separate amplifier is needed in order to exploit the negative Miller effect, this second amplifier can also be used as one of the amplifying stages in a multistage amplifier. A limiting amplifier making use of this principle will be presented in Sect. 6.5.2.

6.3.3 *Other Techniques*

Many other broadband techniques have been presented in literature to increase the *GBW* of an amplifier. Some of these are listed below:

- **Inductive peaking**
  The output capacitance of an amplifier can be partly tuned out by means of an inductor. This leads to *shunt peaking* or *series peaking* if the inductor is placed in parallel or in series with the output node, respectively. Both techniques have been applied in [Wei06] to realize a limiting amplifier in 90 nm CMOS with a voltage gain of 19 dB and a bandwidth of 44 GHz and in [Lia06] to realize an AGC amplifier in 180 nm CMOS for 10 Gbit/s ethernet. If the chip area is limited, the inductive effect can also be obtained by means of active inductors, as in [Säc00] where a limiting amplifier in 250 nm CMOS is presented with a power consumption of 53 mW. The achieved bandwidth is 3 GHz while the gain equals 32 dB.
- **Cherry-Hooper amplifier**
  The *GBW* of a simple differential pair can be increased by realizing the voltage gain in two stages: a transconductance stage and a transimpedance stage. This results in the ***Cherry-Hooper amplifier*** which was first presented in [Che63]. A modified version of the Cherry-Hooper amplifier has been used in [Her05] to implement a limiting amplifier in 180 nm CMOS which shows a voltage gain of 27.4 dB and features bit rates up to 3.5 Gbit/s. The power consumption of this limiting amplifier is 34.2 mW.
- **Active feedback**
  Active negative feedback can be applied as a means of improving the *GBW* of amplifiers. Such an arrangement uses two transconductance stages in series and a third transconductance stage to return a fraction of the output signal to the input of the second stage. It has been shown in [Gal03] that this technique allows one to increase the *GBW* beyond the f_t of the technology. A comparable technique with third-order interleaving active feedback has been applied in [Hua07] to realize a limiting amplifier in 180 nm CMOS with a gain of 42 dB and a bandwidth of 9 GHz. The main drawback of this technique is the high power consumption, namely 189 mW for this implementation, which is needed to design a stable feedback system operating at high speeds.
- **Distributed amplification**
  Distributed amplifiers are a class of amplifiers that do not suffer from the classic trade-off between gain and bandwidth. As explained in [Mac10], signal delay can be exchanged for gain by distributing gain cells along artificial transmission lines. This concept has been applied in [Gal04] to implement an amplifier in 180 nm CMOS with a gain of 15 dB, a bandwidth of 22 GHz, and a power consumption of 190 mW.

6.4 Offset Compensation

It has been shown in Sect. 6.2 that an offset voltage degrades the performance of a limiting amplifier and of the optical receiver it is part of significantly. Although the input offset voltage of a CMOS limiting amplifier is typically around 10 mV, it should be below 1 mV in order not to jeopardize the BER of the optical receiver [Säc05]. An offset compensation circuit is therefore required so as to reduce the input offset voltage of a CMOS limiting amplifier.

Two main classes of offset compensation circuits are generally available. The first class comprises *sampled* solutions while the others can be collected under the term *continuous-time*. The sampled techniques normally use two clock phases. The offset voltage is measured in the first clock phase and the measured offset voltage is subtracted from the signal in the second clock phase. The *auto-zeroing* technique is one of the possibilities to implement this kind of functionality. This principle has been used in [Wit09] where an instrumentation amplifier is presented with a very low input offset voltage of 2.5 μV. Another example in the sampled class is the *correlated double sampling* technique which has been applied in [Saf08] to obtain an input offset current of only 2.4 nA. *Chopper stabilization* is a third manner to solve the offset issue in discrete time steps [Wit07]. This technique can be applied to reduce the offset voltage as well as the pink noise in multipliers, VGAs and mixers, as explained thoroughly in [God08]. In order not to introduce *aliasing*, the signal should be sampled at a frequency that is at least equal to the *Nyquist rate*, which is 2 times the highest frequency component of the signal. This is a problem especially for a wideband amplifier because the bandwidth of such an amplifier, and consequently also that of the accompanying signals, is typically limited by the intrinsic speed of the technology. Sampling the signal at a frequency which is at least 2 times higher than that is not possible because the technology is simply not able to operate that fast.

The other class of offset compensation circuits comprises continuous-time solutions in which a low-pass filter is typically applied to distinguish the offset voltage, which is a DC signal, from the wanted signal. The schematic diagram of a limiting amplifier with this kind of offset compensation is shown in Fig. 6.12a. The DC output voltage is measured by means of the low-pass filter R_{OC}-C_{OC} and amplified by A_{OC}. This amplified DC output voltage is then subtracted from the input voltage by applying it to the inverting input of the differential limiting amplifier. The transfer function of this closed-loop amplifier can be described with the following formula:

$$\frac{v_{out}}{v_{in}} = A_{LA}\frac{1 + j2\pi f R_{OC}C_{OC}}{1 + A_{LA}A_{OC} + j2\pi f R_{OC}C_{OC}}, \tag{6.35}$$

where A_{LA} respresents the small-signal voltage gain of the limiting amplifier. It is assumed to be significantly larger than 1. As can be noticed in the magnitude plot of this transfer function as shown in Fig. 6.12b, the voltage gain at DC of the closed-loop amplifier is inversely proportional to A_{OC}. As the offset voltage at the input, $v_{os,in}$, and at the output, $v_{os,out}$, of the limiting amplifier are DC signals, they are both affected by the feedback loop as follows:

$$|v_{os,in,OC}| = \frac{|v_{os,in}|}{A_{OC}A_{LA}}, \tag{6.36}$$

$$|v_{os,out,OC}| = \frac{|v_{os,out}|}{A_{OC}A_{LA}}, \tag{6.37}$$

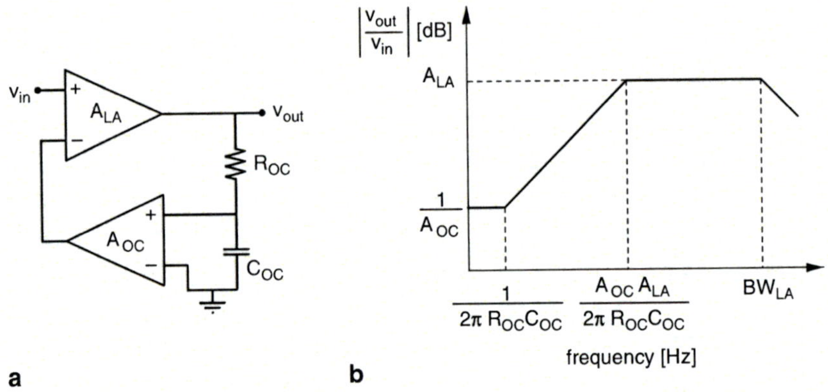

Fig. 6.12 A limiting amplifier with a voltage gain A_{LA} and a bandwidth BW_{LA} around which a continuous-time offset compensation circuit is placed; it consists of a low-pass filter and a voltage amplifier with a voltage gain A_{OC}. **a** Schematic diagram. **b** Magnitude plot of the transfer function

where $v_{os,in,OC}$ and $v_{os,out,OC}$ are the input and output offset voltages of the limiting amplifier with offset compensation, respectively. It can be seen that the offset voltages at the input and the output are reduced by the loop gain. In reality, however, the feedback voltage amplifier also has an offset voltage which has not been considered until now. If its input offset voltage is represented by $v_{os,in,1}$, (6.36) should be rewritten as follows [Säc05]:

$$|v_{os,in,OC}| = \sqrt{\left(\frac{v_{os,in}}{A_{OC}A_{LA}}\right)^2 + \left(\frac{v_{os,in,1}}{A_{LA}}\right)^2}, \tag{6.38}$$

where it is assumed that $v_{os,in}$ and $v_{os,in,1}$ are statistically independent. It can thus be concluded that the offset voltage of the limiting amplifier is suppressed by the gain of both amplifiers, while the offset voltage of the feedback amplifier is only suppressed by the gain of the limiting amplifier. Both voltage gains should be large in order to obtain the minimal resulting offset voltage.

A second effect of the offset compensation feedback loop in Fig. 6.12a is that the frequency response of the resulting amplifier exhibits a high-pass characteristic, as can be seen in Fig. 6.12b. It has been concluded in Sect. 2.5.1 that such a characteristic introduces baseline wander. In order to keep this effect under control, the following condition should be fulfilled:

$$\frac{A_{OC}A_{LA}}{2\pi R_{OC}C_{OC}} < \frac{1}{(2^m - 1)\,T_b}, \tag{6.39}$$

where T_b is the bit period and m is the maximum run length of the applied PRBS signal. Almost all parameters in this formula have already been determined: a minimal

A_{LA} is required to amplify the smallest signals to a rail-to-rail level, A_{OC} is determined by the required suppression of the offset voltage and m and T_b are properties of the received data signal. If (6.39) is not initially satisfied, the only solution is to increase the time constant $R_{OC}C_{OC}$. If a gain of 30 dB is needed for the limiting amplifier and if A_{OC} needs to be at least 20 dB so as to obtain the wanted offset voltage, the required time constant is 6.45 μs for a bit period of 1 ns and a maximum run length of 7. Even if R_{OC} has a resistance of 100 kΩ, the required capacitance is at least 64.5 pF. Such large passive components typically require a large chip area which is clearly a disadvantage from a cost point of view.

The same time constant can also be realized by means of an active low-pass filter, as shown in Fig. 6.13. Instead of being used to reduce the resulting offset voltage, the extra amplifier with voltage gain A_{OC} can also be applied to increase the effective time constant of the low-pass filter by employing the Miller effect on C_{OC}. This poses a trade-off between the resulting offset voltage and the required time constant $R_{OC}C_{OC}$. As the circuit in Fig. 6.13 is in a unity feedback configuration at low frequencies, the resulting input offset voltage is now given by the following formula:

$$|v_{os,in,OC}| = \frac{|v_{os,in}|}{A_{LA}}. \tag{6.40}$$

Although this is A_{OC} times larger than in the configuration of Fig. 6.12a, it is typically smaller than 1 mV. In order to keep baseline wander under control, the time constant $R_{OC}C_{OC}$ is now subjected to the following condition:

$$\frac{A_{LA}}{2\pi\,(1 + A_{OC})\,R_{OC}C_{OC}} < \frac{1}{(2^m - 1)\,T_b}. \tag{6.41}$$

By replacing the passive low-pass filter of Fig. 6.12a with the active implementation of Fig. 6.13, the resulting offset voltages at the input and at the output increase by a factor A_{OC} while the required time constant $R_{OC}C_{OC}$ decreases by a factor A_{OC}^2 for the same amount of baseline wander. Obviously, this is very attractive from an area and cost point of view because the needed chip area to implement this time constant scales with A_{OC}^2 also.

The circuit in Fig. 6.13 contains two feedback loops. Therefore, the analysis of the stability of this circuit is verified by considering two loop gains. The first loop gain is analyzed by cutting the loop around the first voltage amplifier and R_{OC}, as shown in Fig. 6.14a, while the second loop gain is analyzed by cutting the loop around the second voltage amplifier and C_{OC}. In order to identify any stability issues, the

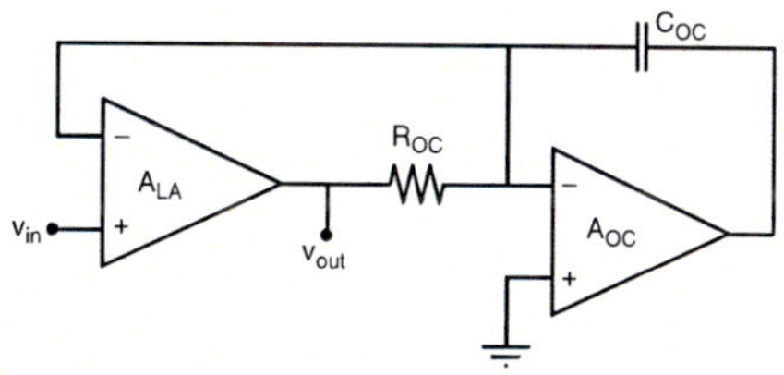

Fig. 6.13 Offset compensation of a limiting amplifier making use of an active low-pass filter

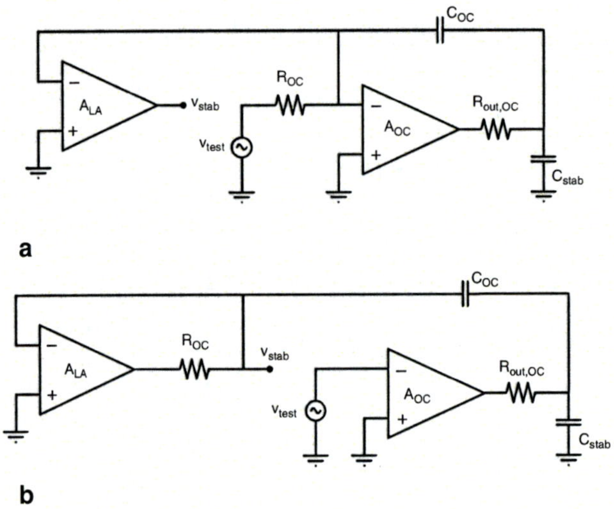

Fig. 6.14 **Stability analysis of a limiting amplifier with active offset compensation.** **a** Schematical diagram to analyze LG_1. **b** Schematical diagram to analyze LG_2

finite bandwidth of the second voltage amplifier is modeled by means of its output resistance $R_{out,OC}$, while C_{stab} has been added to force the circuit to be stable, as will become clear later. The first loop gain can be calculated with the following formula:

$$LG_1 \approx -A_{LA} \frac{1 + j2\pi f R_{out,OC} C_{OC}}{(1 + j2\pi f A_{OC} R_{OC} C_{OC}) \left(1 + j2\pi f \frac{R_{out,OC} C_{stab}}{A_{OC}}\right)}, \tag{6.42}$$

where it is assumed that R_{OC} and C_{OC} are much larger than $R_{out,OC}$ and C_{stab}, respectively, and that A_{OC} is significantly larger than 1. It can be noticed that the loop gain at low frequencies equals $-A_{LA}$. The minus sign, or phase shift of $-180°$, comes from the inverting character of the loop at these frequencies which is needed to obtain stable operation. The two poles and one zero in (6.42) result in a loop gain which eventually evolves to 0 and a phase that asymptotically approaches $-270°$. This can be seen in the Bode plot of LG_1 in Fig. 6.15. The locations of the zero and the two poles in this loop gain can be calculated with the following formulas:

$$f_{z,LG_1} = \frac{1}{2\pi R_{out,OC} C_{OC}}, \tag{6.43}$$

$$f_{p,1,LG_1} = \frac{1}{2\pi A_{OC} R_{OC} C_{OC}}, \tag{6.44}$$

$$f_{p,2,LG_1} = \frac{A_{OC}}{2\pi R_{out,OC} C_{stab}}. \tag{6.45}$$

For frequencies between the zero and the second pole, the magnitude of this loop gain is constant and smaller than 0 dB because R_{OC} is generally much larger than

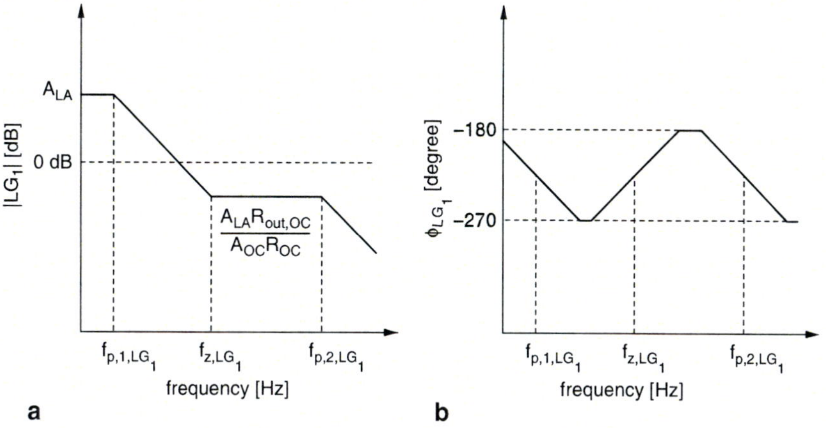

Fig. 6.15 Bode plot of LG_1 of a limiting amplifier with active offset compensation. **a** Magnitude. **b** Phase

$R_{out,OC}$ while A_{OC} and A_{LA} are more alike. Consequently, the stability of this loop is guaranteed.

The second loop gain can be calculated with Fig. 6.14b as follows:

$$LG_2 \approx -\frac{A_{OC}}{A_{LA}} \frac{j2\pi f R_{OC} C_{OC}}{\left(1 + j2\pi f \left(\frac{R_{OC}}{A_{LA}} + R_{out,OC}\right) C_{OC}\right)\left(1 + j2\pi f \frac{R_{OC} R_{out,OC} C_{stab}}{R_{OC} + A_{LA} R_{out,OC}}\right)}, \tag{6.46}$$

where the same assumptions have been made as before. A zero is now located at DC while the two poles finally reduce this loop gain to 0. As can be derived from this formula, the locations of both poles are determined by the following formulas:

$$f_{p,1,LG_2} = \frac{1}{2\pi \left(\frac{R_{OC}}{A_{LA}} + R_{out,OC}\right) C_{OC}}, \tag{6.47}$$

$$f_{p,2,LG_2} = \frac{R_{OC} + A_{LA} R_{out,OC}}{2\pi R_{OC} R_{out,OC} C_{stab}}. \tag{6.48}$$

The Bode plot of LG_2 is shown in Fig. 6.16. A constant magnitude can now be noticed in between both poles. However, in contrast to what was the case with LG_1, this constant magnitude is larger than 1 for typical parameter values, as can be seen in the following inequality:

$$\frac{A_{OC} R_{OC}}{R_{OC} + A_{LA} R_{out,OC}} > 1 \Longleftrightarrow R_{out,OC} < \frac{A_{OC} - 1}{A_{LA}} R_{OC}, \tag{6.49}$$

which is always the case for the assumptions that have been made before, namely $R_{OC} \gg R_{out,OC}$ and $A_{OC} \approx A_{LA}$. At first sight, there seems to be no stability issue for

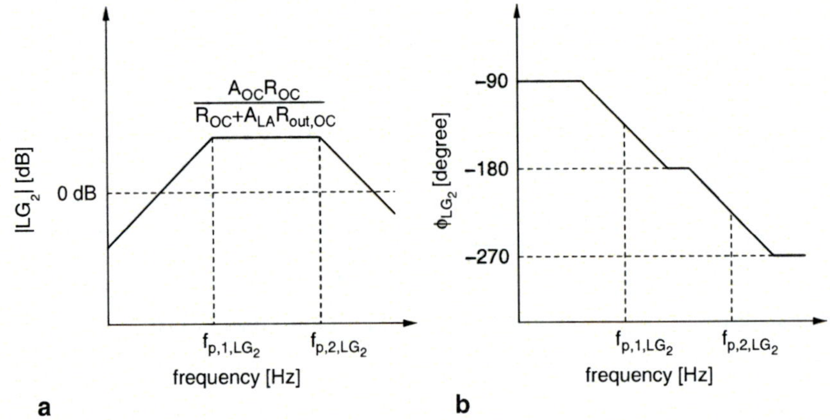

Fig. 6.16 Bode plot of LG_2 of a limiting amplifier with active offset compensation. **a** Magnitude. **b** Phase

this feedback loop either. However, in reality the first voltage amplifier with gain A_{LA} also has a finite bandwidth BW_{LA} which has not been taken into account in (6.46). Therefore, the system can become unstable if the second zero crossing of LG_2 is not located at a frequency which is at least 3 times lower than that bandwidth. The following stability condition can consequently be prescribed:

$$3 \cdot \frac{A_{OC}}{2\pi R_{out,OC} C_{stab}} < BW_{LA}. \tag{6.50}$$

From this condition, it becomes clear why capacitor C_{stab} has been added to the circuit as it is needed to guarantee stable operation.

6.5 Post Amplifier Designs

Two circuit designs of a limiting amplifier and their respective simulation results are shown below in order to illustrate the principles that have been introduced in the previous sections. While the first limiting amplifier has been designed in a 130 nm CMOS technology, the second limiting amplifier has been designed in a 90 nm CMOS technology and applies the principles of the negative Miller effect to obtain a high bandwidth and active offset compensation, resulting in a very small chip area.

6.5.1 A 4-stage Limiting Amplifier

This design is intended to be used in collaboration with the differential TIA that has been discussed in Sect. 4.4.1. The simulated bandwidth of this TIA is 2.71 GHz or

2.85 GHz, depending on the biasing condition. Consequently, the maximum bit rate that can be received properly by it is around 4.5 Gbit/s because it has been derived in Sect. 2.5.2 that a bandwidth of 0.6–0.7 · B Hz is required if the bit rate equals B bit/s. For the same bit rate, the limiting amplifier needs to have a small-signal bandwidth of 1–1.2 · B Hz in order to keep the group delay variation under control, as concluded in Sect. 6.2. As a result, the required bandwidth of the limiting amplifier is 5 GHz or higher. Thanks to the large transimpedance gain of 4.95 kΩ and due to the input noise current of 624 nA_{RMS} and 669 nA_{RMS} of the two TIAs in Sect. 4.4.1, the input signal swing of the limiting amplifier is larger than 10 mV if a BER below 10^{-12} is aimed at. Therefore, the limiting amplifier does not need to have a very large small-signal gain. A value of 30 dB is targeted in this design.

The circuit diagram of the presented limiting amplifier is shown in Fig. 6.17. The gain is provided by means of four cascaded stages because it has been shown in Sect. 6.3.1 that this is optimal in terms of the required *GBW* of a single stage, as can be concluded from Fig. 6.7 as well as for the total power consumption which follows from Fig. 6.8. The first two amplifier stages are realized by means of a differential amplifier with nMOS input transistors. Linear pMOS transistors are used as loads because they have a very small physical size and consequently exhibit a very small parasitic capacitance to the substrate. By contrast, the latter two stages use polysilicon resistors instead which behave more linearly for the larger swings that are present there.

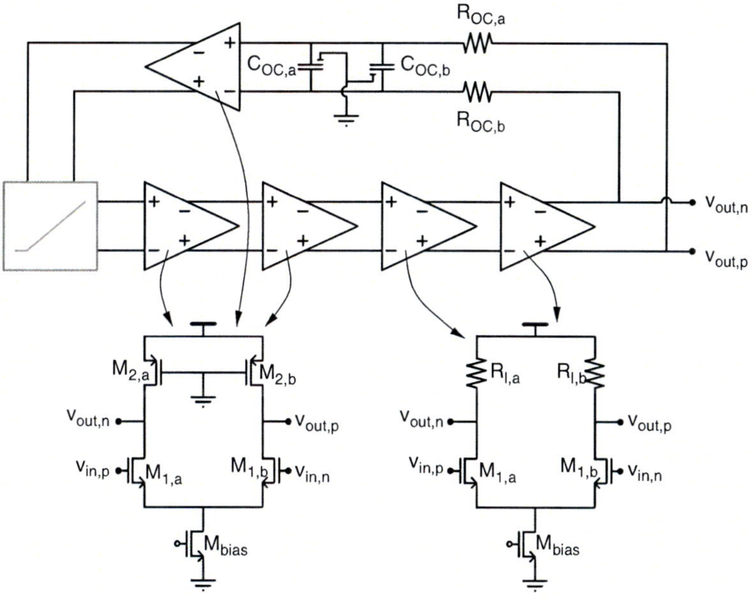

Fig. 6.17 Circuit diagram of the 4-stage limiting amplifier in 130 nm CMOS in which the offset compensation feedback loop is closed in the equalizer of Sect. 5.3.1

The offset compensation circuit suppresses not only the offset voltage of the limiting amplifier but also that of the equalizer. This has been achieved by closing the feedback loop in the equalizer which is feasible because this limiting amplifier is intended to be used with the equalizer of Sect. 5.3.1 where indeed two inputs are available to regulate the output offset voltage. Two low-pass filters are needed to obtain the output offset voltage due to the differential topology. In [Gal03, Lia06, Yoo07, Hua07], two resistors and two capacitors to ground have been applied to realize these filters. By contrast, two resistors, $R_{OC,a}$ and $R_{OC,b}$, and only a single differential capacitor have been used in the limiting amplifier presented here. To obtain the most symmetrical behavior, the capacitor is divided in two equal parts, $C_{OC,a}$ and $C_{OC,b}$, so that each side has the same parasitic capacitance to ground. Compared to placing two capacitors to ground, placing only a single differential capacitor has two obvious advantages:

1. Only a quarter of the capacitance is needed to realize the same time constant because only a single capacitor is needed instead of two and because that capacitor can be 2 times smaller due to its differential placement;
2. Common-mode signals that would appear at the output of the limiting amplifier are rejected over the entire frequency range because these kinds of signals are not filtered by the differential low-pass filter.

The amplifier in the offset compensation feedback path has the same topology as the first two stages of the limiting amplifier and is used to reduce the resulting offset voltage even more by its gain.

This limiting amplifier has been designed in the same 130 nm CMOS technology which has been used for the differential TIA of Sect. 4.4.1 and the equalizer of Sect. 5.3.1. All differential amplifying stages conduct an identical current, namely 828 μA. The transconductance of the input transistors is consequently also the same for all the stages, namely 3.29 mS for an overdrive voltage of 183 mV. However, the voltage gain of the first two stages is slightly larger than that of the latter two stages due to a higher output resistance. While the first two stages realize a voltage gain of 9.03 dB over a bandwidth of 10.6 GHz, the voltage gain and bandwidth of the last two stages are 6.95 dB and 17.8 GHz, respectively. As a result, the total gain of the limiting amplifier is 31.9 dB and the bandwidth of the cascade is 4.74 GHz. This can be verified on the simulated Bode plot which is shown in Fig. 6.18. DC signals, including the offset voltage, are transferred to the output with an attenuation of −11.8 dB. The value of $R_{OC,a}$ and $R_{OC,b}$ is 20 kΩ and that of $C_{OC,a}$ and $C_{OC,b}$ is 52.5 pF. This results in a lower cut-off frequency of the offset compensation feedback loop, which is at 6.31 MHz. The total current consumption of the four amplifying stages and the feedback amplifier is 5.04 mA. Together with a supply voltage of 1.2 V this results in a power consumption of only 6.05 mW. The performance of the limiting amplifier presented here is summarized in Table 6.2.

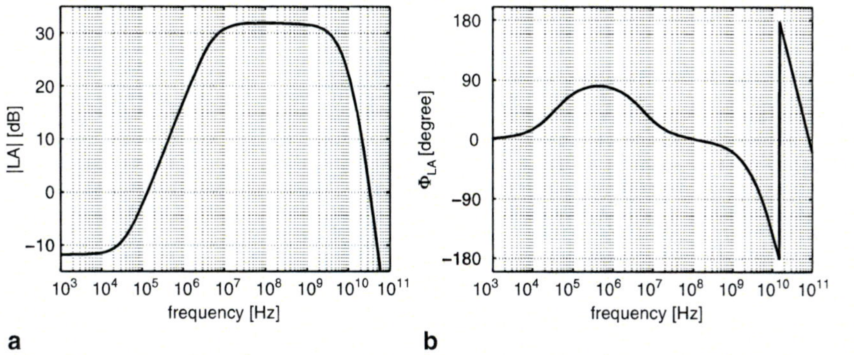

Fig. 6.18 Simulated Bode plot of the 4-stage limiting amplifier with an offset compensation feedback loop in 130 nm CMOS. **a** Magnitude. **b** Phase

Table 6.2 Overview of the simulated performance figures of the 4-stage limiting amplifier with offset compensation in 130 nm CMOS

AC gain	31.9 dB
DC gain	−11.8 dB
bandwidth	4.74 GHz
lower cut-off frequency	6.31 MHz
current consumption	5.04 mA

6.5.2 A Limiting Amplifier with Negative Miller Capacitors and Active Offset Compensation

The objective of this design of which a schematical diagram is shown in Fig. 6.19 is to demonstrate the effectiveness of the negative Miller effect to realize a system with high bandwidth and an active offset compensation feedback loop to reduce the time constant which has to be realized with passive components. By contrast with the previous design, this limiting amplifier is meant to be used as a separate building block. Therefore, the input nodes are terminated with the same resistance as the characteristic resistance of the measurement setup, 50 Ω, so as not to introduce any reflections at the input side. The same is done at the output side by means of the output buffer, which is not discussed here. As can be seen in Fig. 6.19, the presented limiting amplifier can be divided into three parts: the input buffer, the amplifier core and the active offset compensation feedback loop.

The input buffer has double functionality, namely termination of the input nodes and subtraction of the measured output offset voltage from the input signal. As can be seen in Fig. 6.19, a *differential difference amplifier* is used as the input buffer. The input termination is performed by means of the 100 Ω resistor which is placed differentially between the gates of the two input transistors $M_{1,a}$ and $M_{1,b}$ so that a 50 Ω resistance to ground is seen at each of the input nodes. The differential difference amplifier also offers a second differential input pair $M_{2,a}$-$M_{2,b}$ which is

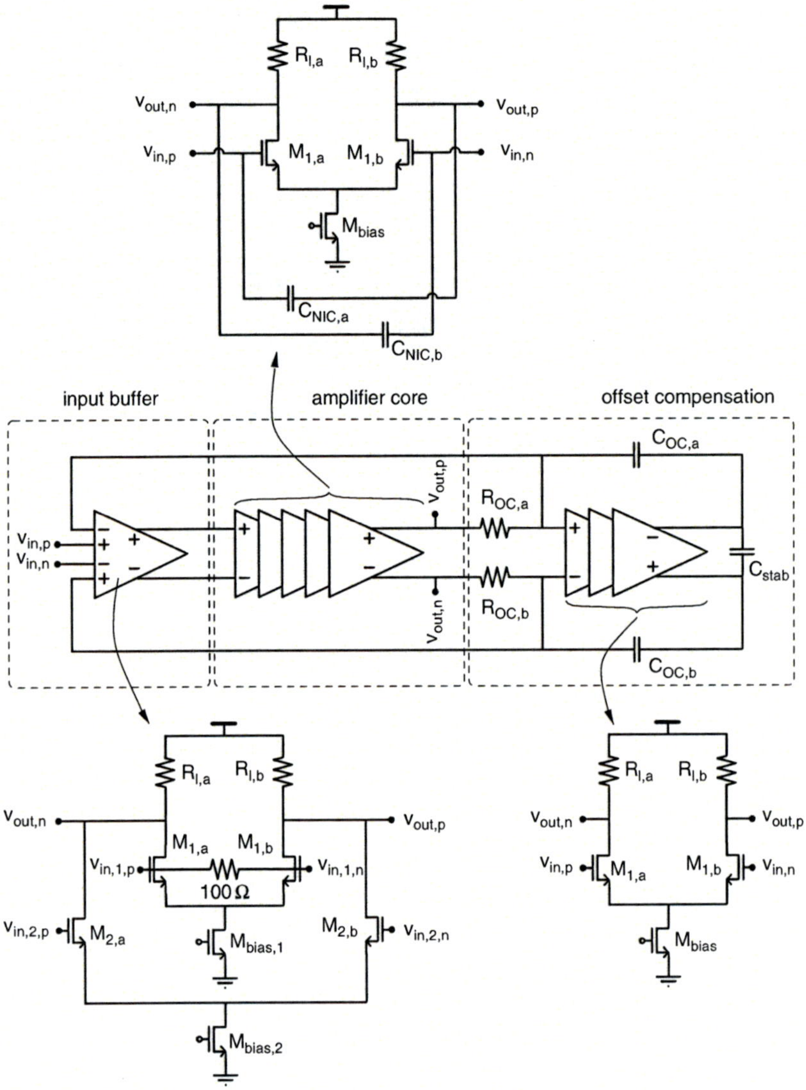

Fig. 6.19 **Circuit diagram of the limiting amplifier with negative Miller capacitors and active offset** compensation in 90 nm CMOS

used to close the offset compensation feedback loop. The input signals of the two differential pairs are summed in the shared load resistors $R_{l,a}$ and $R_{l,b}$.

The amplifier core is composed of a cascade of five amplifier stages which all share the same topology. As can be seen in Fig. 6.19, the gain is realized by means of a resistively loaded differential pair with nMOS input transistors. Capacitors $C_{NIC,a}$ and $C_{NIC,b}$ are placed between the input and output nodes with the same polarity so that the negative Miller effect is invoked. Because of this, the effective input capacitance of the stage decreases while its output capacitance increases, as explained in Sect. 6.3.2. The smaller input capacitance results in a pole on the output node of the preceding stage which is at a higher frequency. The larger output capacitance is compensated in its turn by the input capacitance of the subsequent stage. Instead of the four stages of the previous design, five stages have been cascaded here because the objective with this design was to realize a voltage gain of at least 35 dB, which is almost 2 times more than in the previously discussed limiting amplifier.

The offset compensation feedback loop uses an active low-pass filter where the time constant is boosted by means of a 3-stage voltage amplifier. Every stage again has the same topology which is also shown in Fig. 6.19. As the bandwidth of this offset compensation amplifier is not so critical as that of the limiting amplifier itself, negative Miller capacitors have not been applied here. Only three stages have been cascaded because too large a gain could possibly saturate the outputs of the offset compensation amplifier if the offset voltage of the limiting amplifier is too large. The voltage gain of this amplifier would collapse accordingly and the effective time constant of the low-pass filter would decrease significantly. Finally, a capacitor C_{stab} has been placed differentially at the output to guarantee stable operation.

The design has been realized in a 90 nm CMOS process with a nominal supply voltage of 1 V. In order to keep the noise to a minimum, the major portion of the available current, namely 4.16 mA, has been reserved for the input buffer. This current has been divided equally between the two parallel differential pairs so that the voltage gain from both input pairs to the output is the same and is approximately equal to 0 dB. The five stages of the limiting amplifier have relative sizes as follows: 3-2-1-1-1. As a result, the current through the first stage is 1.5 times larger than that through the second stage and 3 times larger than that through the latter three stages. Firstly, this is advantageous for the capacitive loading of the first and second stages, optimizing the poles on their output nodes. Secondly, it is optimal in terms of noise as the first stage produces the dominant noise contribution. The total current through the five amplifier stages is 5.54 mA. Every stage realizes the same voltage gain of 7.46 dB. From the first to the last stage, $C_{NIC,a}$ and $C_{NIC,b}$ have the following sizes: 24 fF—27 fF—21.5 fF—14.5 fF—11.5 fF. The current of the offset compensation amplifier has been distributed over its three stages according to the following ratio: 8-1-1. This is again beneficial for the noise contribution of this part of the circuit. More importantly, however, the offset voltage of the offset compensation amplifier itself is minimized by this choice because the first stage, which mainly determines the offset voltage, occupies the largest area. Moreover, in order to reduce the offset voltage of this amplifier even more the lengths of the input transistors of all three stages equal 200 nm, 2.5 times greater than the minimal length. The voltage gain of the offset compensation amplifier is 22.2 dB. Together with a resistance of 100 kΩ for $R_{OC,a}$ and $R_{OC,b}$ and a capacitance of 10 pF for $C_{OC,a}$ and $C_{OC,b}$, the resulting lower cut-off frequency is 925 kHz. The total current consumption of the offset compensation amplifier is 2.96 mA. The total current consumption of the limiting amplifier is 12.7 mA, resulting in a power consumption of 12.7 mW for a supply voltage of 1 V. As can be seen in Fig. 6.20, the voltage gain after parasitic extraction

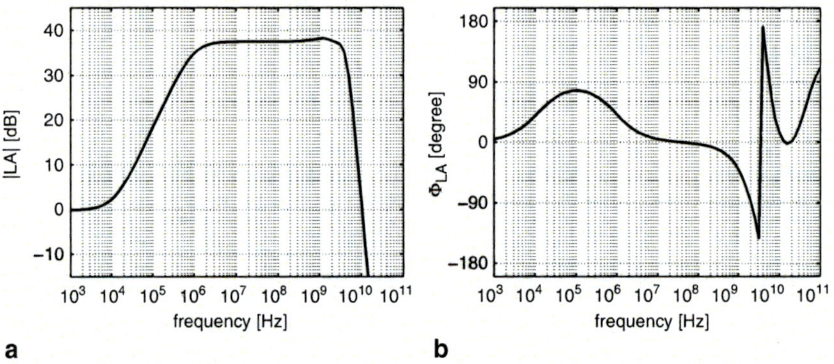

Fig. 6.20 **Bode plot of the limiting amplifier with active offset compensation and negative Miller** capacitors in 90 nm CMOS after parasitic extraction. **a** Magnitude. **b** Phase

Table 6.3 Overview of the performance of the limiting amplifier with active offset compensation and negative Miller capacitors in 90 nm CMOS

AC gain	37.5 dB
DC gain	0 dB
bandwidth	4.15 GHz
lower cut-off frequency	925 kHz
RMS input noise	285 μV_{RMS}
current consumption	12.7 mA

is 37.5 dB and the bandwidth of the system is 4.15 GHz. The simulated RMS input noise is 285 μV_{RMS} which results in very high sensitivity as will be shown in the next chapter. The performance of this design is summarized in Table 6.3.

6.6 Conclusion

The output signals of the TIA and the equalizer have swings which are too low to be considered as digital data. Therefore, the signal swing needs to be increased significantly which is the task of the post amplifier.

The difference between a limiting amplifier and an AGC amplifier is examined first. It is shown that a limiting amplifier has a superior performance in the intended application, namely in an optical receiver chain processing a digital data stream where linearity is only of minor importance. Only this architecture is consequently considered in the remainder of this chapter.

The most important specifications of a post amplifier, and a limiting amplifier in particular, are discussed thereafter. The small-signal voltage gain needs to reach a certain minimal value in order not to introduce a noticeable power penalty in the decision circuit. It is shown that the small-signal bandwidth needs to be $1 - 1.2 \cdot B$ Hz if the bit rate equals B bit/s so as to keep the group delay variation within 10% of the bit period. The offset voltage is another important specification because it can degrade the BER of the receiver significantly. It is concluded that an offset compensation circuit is necessary in a CMOS limiting amplifier.

Next, some broadband design techniques are presented. A multistage topology enables the realization of an amplifier with a very large *GBW* and low power consumption. Another design technique is the NIC. By means of a negative resistance and a negative capacitance, this converter can increase the output resistance of an amplifier and decrease its output capacitance so that the gain as well as the bandwidth are enhanced. The negative Miller effect can be applied to realize a negative capacitance.

Subsequently, the functionality of the offset compensation circuit is treated. Because this circuit demands the integration of a large resistance and capacitance, an active technique is presented to boost the time constant of a low-pass filter. It is shown how such a topology can be stabilized.

Finally, two circuit designs are shown. The first design is designed in a 130 nm CMOS process and uses four amplifying stages to realize a gain of 31.9 dB over a bandwidth of 4.74 GHz. The second design is used as a demonstrator to illustrate the effectiveness of the NIC and the active offset compensation circuit in order to obtain a limiting amplifier with low power consumption and a small physical area.

Chapter 7
Chip Implementations

All of the analog building blocks of an optical receiver front-end have been treated extensively in the previous chapters. The realized chip implementations making use of a combination of some of these building blocks are discussed in this chapter.

In order to be able to compare objectively the realized chip implementations to the ones that have been published recently in open literature, a new *figure of merit* (FOM) is introduced in the first section.

The first chip implementation has been integrated in a 130 nm CMOS technology. A differential photodiode with a diameter of 60 μm, a differential TIA, an optimized equalizer and a limiting amplifier have been collected on a single chip which transforms the small optical input signal into a rail-to-rail output voltage. Bit rates up to 4.5 Gbit/s can be received properly by this receiver at a BER below 10^{-12} if the optical input power is at least −3.8 dBm or −3.4 dBm, depending on the biasing condition.

The second implementation is a modified version of the first chip where the differential photodiode has been replaced by a speed-enhanced photodiode with a diameter of 60 μm and where the differential TIA has been optimized for this new photodetector. Signals with a bit rate up to 5.5 Gbit/s can be processed by this chip at a BER below 10^{-12} if the optical input power is at least −3.4 dBm.

The third chip implementation has been integrated in a 180 nm CMOS process and is intended to be applied as a POF receiver. Therefore, the integrated photodiode has a diameter of 1 mm. Optical signals with a bit rate up to 300 Mbit/s and 500 Mbit/s can be processed reliably at a supply voltage of 1.12 V and 1.42 V respectively.

The fourth chip has been integrated in a 90 nm CMOS technology. In contrast with the other implementations only a limiting amplifier is integrated on this chip which can be considered as a demonstrator of some of the design techniques that have been presented in Chap. 6. The proposed circuit has a small-signal gain of 35 dB and a bandwidth of 4.15 GHz. The input sensitivity for a BER of 10^{-12} is 2.75 mV, 2.9 mV and 3.75 mV for a bit rate of 3, 4 and 5 Gbit/s, respectively.

F. Tavernier, M. Steyaert, *High-Speed Optical Receivers with Integrated Photodiode in Nanoscale CMOS,* Analog Circuits and Signal Processing 5,
DOI 10.1007/978-1-4419-9925-2_7, © Springer Science+Business Media, LLC 2011

7.1 A New FOM for Integrated Optical Receivers

A FOM is required to compare different chip implementations of optical receivers in an objective manner. Due to the specific nature of the issues that play in CMOS integrated photodetectors, as discussed thoroughly in Chap. 3, the following conditions should be fulfilled by an implementation that is to be qualified by this FOM:

1. The photodiode must be integrated together with the circuitry in a silicon CMOS process so that the implementation can be considered as a single-chip solution;
2. No extra processing steps should be required which could increase the performance of the photodiode artificially;
3. The measurements should be made with an optical input signal so that the real life performance is mimicked as closely as possible;
4. Modulated light with a wavelength of 850 nm should be used as the input signal;
5. The circuit should have a completely differential nature in order not to favor single-ended implementations which have a lower power consumption;

While the former four rules are imposed to take care of specific issues related to CMOS photodiodes, the fifth rule is added because the drawbacks of a single-ended implementation, for example a higher sensitivity to fluctuations of the supply voltage, are not taken into account in the proposed FOM:

$$\text{FOM} = \frac{\text{bit rate [Gbit/s]} \cdot |\log \text{BER}| \cdot |P_s \text{ [dBm]}| \cdot \text{gain [dB}\Omega] \cdot \text{PD diameter}^2 \text{ [}\mu\text{m}^2]}{\text{power consumption [mW]} \cdot L_{min}^2 \text{ [nm}^2]} \tag{7.1}$$

First of all, a high bit rate results in a high FOM. As discussed in Sect. 2.5.3 the BER and the sensitivity P_s are related because a higher optical input power is needed to obtain a lower BER. On a logarithmic scale this relation has a more or less linear behavior as can be seen in the Q-function in Fig. 2.19. Therefore, both are collected in the numerator of the FOM and multiplied together. Some implementations of optical receivers do not provide a post amplifier and are consequently not able to deliver a signal with an adequate signal swing to the CDR circuit. Therefore, the total gain from the photodiode current to the output voltage is also included in the numerator. Having a photodiode with a large diameter is advantageous from a system point of view because it facilitates the assembly of the system. As a result, it is squared in the numerator of the FOM because the photodiode area defines the parasitic capacitance and consequently the performance of the complete receiver. The first measure in the denominator is the total power consumption which obviously should be as low as possible. The second measure in the denominator is L_{min}, the minimal linewidth of the technology, which determines the performance of the photodiode to a great extent for two reasons. Firstly, the smaller the minimal linewidth the more interconnect metal layers are typically stacked and this negatively affects the optical transmission coefficient, as concluded in Sect. 3.2.4. Secondly, the smaller the minimal linewidth the higher the applied doping concentrations which result in a shallower SCR between the n-well and p-substrate of an n-well/p-sub photodiode. Due to this dual effect the minimal linewidth is squared in the FOM.

7.2 A 4.5-Gbit/s Optical Receiver with an Integrated Photodiode in 130-nm CMOS

This implementation is intended to be applied as the analog front-end of an optical communication system making use of a MMF with a core diameter of 50 μm as the communication medium transporting modulated light with a wavelength of 850 nm. The PRBS which is used to evaluate the performance of this chip has a length of $2^7 - 1$. This implementation is focused mainly on a high bit rate and, to a lesser extent, on low power consumption and high sensitivity.

7.2.1 The Chip

The chip architecture can be seen in Fig. 7.1. It consists of a differential photodiode, a differential TIA, an optimized differential equalizer, a 4-stage limiting amplifier with offset compensation and an output buffer. All building blocks, including the photodiode, are integrated on the same chip so that a complete optical receiver is realized on a single chip.

The photodiode The diameter of the integrated photodiode equals 60 μm in order to collect all of the light leaving the MMF with a core diameter of 50 μm. An n-well/p-sub junction has been selected because this type of junction offers the widest

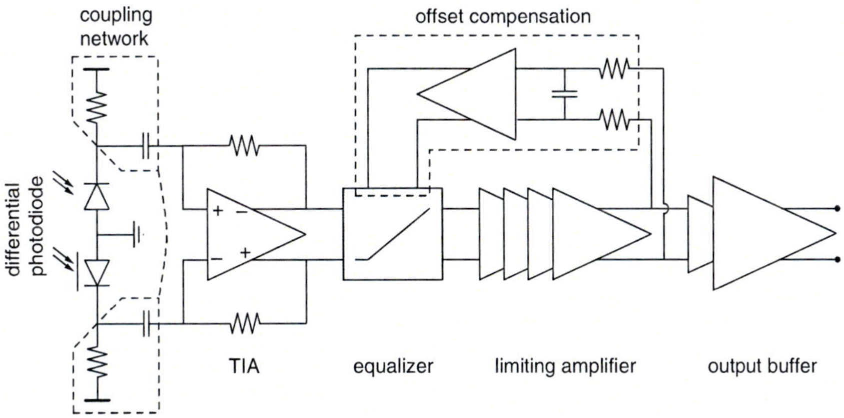

Fig. 7.1 Architecture of the 4.5 Gbit/s optical receiver with an integrated photodiode in 130 nm CMOS

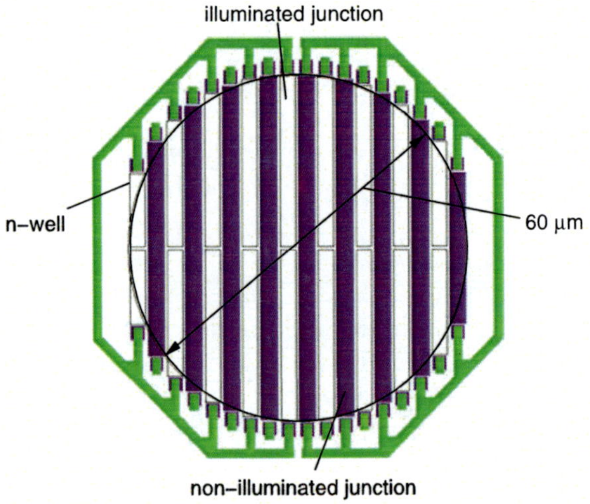

Fig. 7.2 Layout of the differential n-well/p-sub photodiode with a diameter of 60 μm

SCR in a standard CMOS technology. Consequently, the speed and responsivity are maximized while the parasitic photodiode capacitance is minimized, as discussed in Chap. 3. Because the main focus of this implementation is on the attainable bit rate, a differential photodiode structure has been chosen. A simplified version of its layout is shown in Fig. 7.2. The alternating pattern of illuminated and non-illuminated junctions can clearly be observed.

The periodicity of the layout equals 6.68 μm. As discussed in Sect. 3.2.2 a smaller periodicity would result in a higher photodiode bandwidth. However, a smaller periodicity would also result in a smaller responsivity because more area would be wasted due to the minimal distance that needs to be maintained between two n-wells as imposed by the manufacturer. Also, if the periodicity were decreased, the parasitic photodiode capacitance would increase because the sidewall capacitance would be larger for the same photodiode area.

The cathode of the illuminated photodiode is formed by means of an octagonal ring in the eighth metal layer on the left side of the photodiode and in the seventh metal layer on the right side. On the contrary, the cathode of the non-illuminated photodiode is connected to the eight metal layer at the left side and to the seventh metal layer on the right side. This has been done to keep the structure as symmetrical as possible. The two top metal layers provided by the technology have been used for this octagonal contact so as to minimize the resistance introduced as well as the capacitance to the substrate.

The simulated parasitic capacitance of the illuminated and non-illuminated photodiodes in Fig. 7.2 equals 1.02 pF for a reverse voltage of 1.2 V, the nominal supply voltage of the 130 nm technology used. In order to be able to apply this maximum reverse voltage over the photodiode, a coupling network has been placed between the photodiode and the TIA, as shown in Fig. 7.1. A resistor of 20 kΩ connects the cathode of the photodiode to the supply voltage. A capacitor of 10 pF couples this cathode to one of the TIA inputs if the frequency is higher than 796 kHz, the pole frequency which is introduced by this coupling network. The signal should consequently contain only frequencies which are higher than this pole. For the intended $2^7 - 1$ PRBS this should not be a problem if the bit rate is high enough. A second effect of the coupling network is a capacitive division of the photocurrent between its 10 pF capacitor and the parasitic photodiode capacitance of 1.02 pF. With these capacitances it can be calculated that almost 91% of the photocurrent flows towards the TIA. The third effect of the coupling network is that the offset between the currents of the illuminated and non-illuminated photodiodes is cancelled before entering the TIA.

The TIA The TIA is the same as the one that has been presented in Sect. 4.4.1. Two design cases have been proposed there which differ only in the sizing and biasing of the first stage of the voltage amplifier. While both designs provide the same transimpedance of 4.95 kΩ and input capacitance of 320 fF, the other specifications are somewhat different:

- **Design 1**
  $BW_{TIA} = 2.85$ GHz, RMS input noise current = 624 nA_{RMS}, current consumption = 77.4 mA.
- **Design 2**
  $BW_{TIA} = 2.71$ GHz, RMS input noise current = 669 nA_{RMS}, current consumption = 31.5 mA.

Both designs have been implemented, resulting in two chips which differ only in the design of the TIA. The measured performance difference between the chips will be presented in Sect. 7.2.3.

In both cases, the gate-source capacitance of the input transistor of the first stage is 320 fF, about 30% of the photodiode capacitance, as can be verified in Table 4.3. This is optimal regarding the RMS input noise current of this shunt-shunt feedback TIA, which was also concluded in Sect. 4.2. A larger input transistor with an accordingly larger transconductance would result in lower noise in this transistor. However, due to the increased total input capacitance, the feedback resistance should be made smaller so as to maintain the same bandwidth. This would reduce the transimpedance and increase the noise. By contrast, a smaller input transistor would generate too much noise on its own.

The Equalizer The equalizer that is used in this implementation is the one that has been presented in Sect. 5.3.1. It is able to compensate the roll-off of a differential CMOS photodiode up to approximately 5 GHz. As can be deduced easily from the circuit diagram in Fig. 5.7, the tuning characteristic of this equalizer can be adapted by altering the gate voltage of the linear nMOS transistors $M_{R_1,a}$-$M_{R_1,b}$ and $M_{R_2,a}$-$M_{R_2,b}$. This feature can be used during operation to optimize the performance of the chip and, equivalently, to minimize the BER. The current consumption of this differential equalizer is 6.39 mA.

Limiting Amplifier and Offset Compensation The 4-stage limiting amplifier with offset compensation described in Sect. 6.5.1 is placed after the equalizer. It is able to deliver a small-signal gain of 31.9 dB over a bandwidth of 4.74 GHz. The offset compensation circuit suppresses the offset voltage of this limiting amplifier by 11.8 dB and thereby leads to a high-pass characteristic with a pole at 6.31 MHz. This cut-off frequency is low enough for the intended $2^7 - 1$ PRBS if the bit rate is at least 801 Mbit/s, as can be calculated with (6.39). This limiting amplifier and offset compensation circuit together consume a current of 5.04 mA.

Output Buffer The output buffer is composed of a set of four cascaded buffer stages with a tapered design so that it presents a small load capacitance to the limiting amplifier while being able to deliver enough current to the 50 Ω input impedance of the measurement equipment. In order to eliminate any reflections at the output side, the small-signal output resistance of the output buffer has been designed with a value of 50 Ω. Ideally, this buffer should not alter the signal which is applied at its input. Therefore, it has a simulated small-signal gain of 0 dB and a bandwidth which is high enough so that it does not compromise the performance of the rest of the building blocks.

7.2.2 Measurement Setup

The most significant measurement of an optical receiver is the quantification of its BER as a function of the bit rate and of the optical input power. This can be done by the measurement setup that is represented schematically in Fig. 7.3. A PRBS with

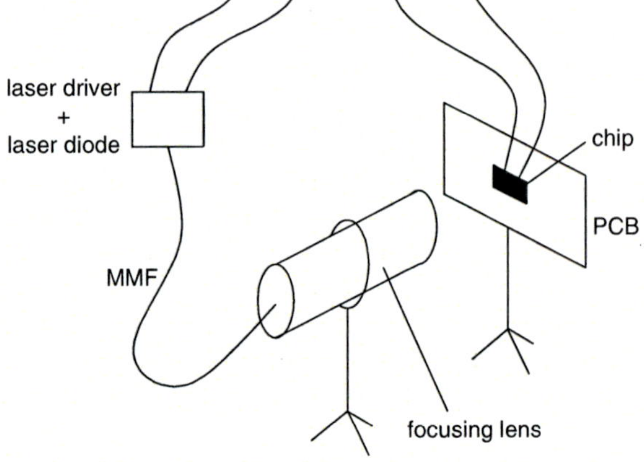

Fig. 7.3 Schematical representation of the measurement setup of the presented optical receiver chip

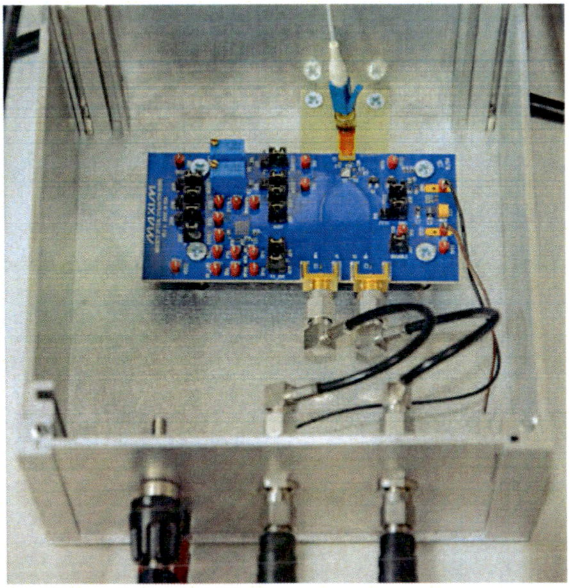

Fig. 7.4 Photograph of the laser driver and laser diode which are both mounted in an aluminum box; the input signals arrive at the bottom while the MMF leaves the box at the top

the chosen length is created by the generator module of the BER tester and applied at the input of a laser driver. This driver transforms its input data into a modulated current which is steered through a laser diode. The light with a center wavelength of 850 nm that is generated by the laser diode is modulated accordingly. The laser driver and laser diode are mounted together inside an aluminum box as can be seen in Fig. 7.4. The data which is now in the optical domain is guided by a MMF with a core diameter of 50 μm and a length of approximately 1 m. A lens is connected at the end of this fiber so that the light can be focused easily on the integrated photodiode[1]. Therefore, the chip is mounted first on an alumina substrate which is placed inside a CuBe box. This box is connected electrically and mechanically with a PCB that provides all necessary currents and voltages to the chip. This can be seen in Fig. 7.5. The output signals of the chip are connected to the inputs of the analyzer module of the BER tester. In order to determine the BER of the chip, the BER tester counts the number of discrepancies between the generated PRBS and the received bit stream.

7.2.3 Measurement Results

The design has been integrated in a 130 nm CMOS process with a nominal supply voltage of 1.2 V. The chip photograph of the optical receiver is shown in Fig. 7.6. Although two chips have been realized which differ only in the design of the TIA,

[1] The focusing lens has a spot size of 50 μm so as to mimic the situation where the MMF with a core diameter of 50 μm is aligned directly on top of the integrated photodiode without using a lens.

Fig. 7.5 Photograph of the chip which is mounted on an alumina substrate inside a CuBe box; the box is connected to the PCB which provides all required currents and voltages; the MMF ending in a focusing lens can be seen in the left part of the picture

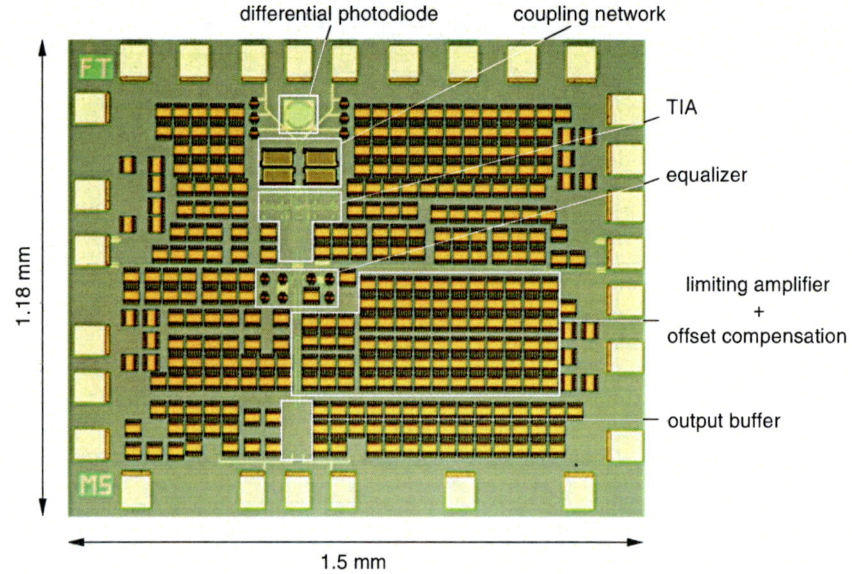

Fig. 7.6 Chip photograph of the 4.5 Gbit/s optical receiver with an integrated photodiode in 130 nm CMOS

only a single chip photograph is shown because both look exactly the same. The chip area, including the biasing circuits, decoupling capacitors and bondpads, equals 1.77 mm^2. Chip 1, the implementation with the first design option for the TIA, has a total current consumption of 115 mA, resulting in a power consumption of 138 mW for a supply voltage of 1.2 V. Chip 2, the implementation with the second TIA of which part of the circuit is biased in weak inversion, has a current and power consumption of 61.8 mA and 74.2 mW respectively, again for a supply voltage of 1.2 V. All these consumption figures have been measured without taking account of the output buffer.

The DC responsivity of the differential photodiode has been assessed by focusing an 850 nm light beam with a known optical power onto the device and measuring the currents of the illuminated and non-illuminated photodiodes. The responsivities of the separate photodiodes are 50.3 mA/W and 45.3 mA/W, respectively, which results in a responsivity of the differential photodiode of only 5 mA/W. This is much lower than has been achieved previously with a differential photodiode in a 180 nm CMOS technology: in [Her07] a value of 30 mA/W is reported while in [Che07b] a value of 70 mA/W has been achieved. As discussed in Chap. 3, this is due to the **higher doping concentrations and the lower optical transmission coefficient of the** 130 nm CMOS technology used.

Some measured eye diagrams of this optical receiver chip are shown in Fig. 7.7 for an optical input power of −3.2 dBm, a PRBS with a length of $2^7 - 1$ and bit rates of 3 Gbit/s and 4.5 Gbit/s. By comparing Fig. 7.7a with Fig. 7.7b, it can be noticed that the eye of chip 1 has a smaller opening for the higher bit rate. The same observation can be made when comparing the eye diagrams of chip 2 in Fig. 7.7c and Fig. 7.7d. When comparing the eye diagrams of both chips at the same bit rate, it can be observed that the eye opening of chip 1 is always larger than that of chip 2. This is the consequence of the slightly lower bandwidth and higher noise of the TIA in chip 2.

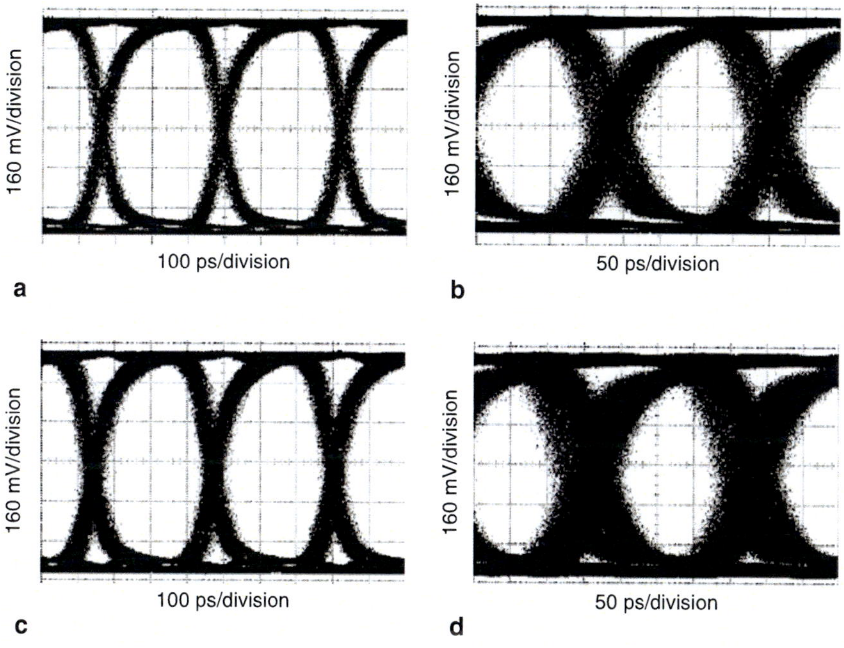

Fig. 7.7 Measured eye diagrams at bit rates of 3 Gbit/s and 4.5 Gbit/s for an optical input power of −3.2 dBm. **a** chip 1; 3 Gbit/s. **b** chip 1; 4.5 Gbit/s. **c** chip 2; 3 Gbit/s. **d** chip 2; 4.5 Gbit/s

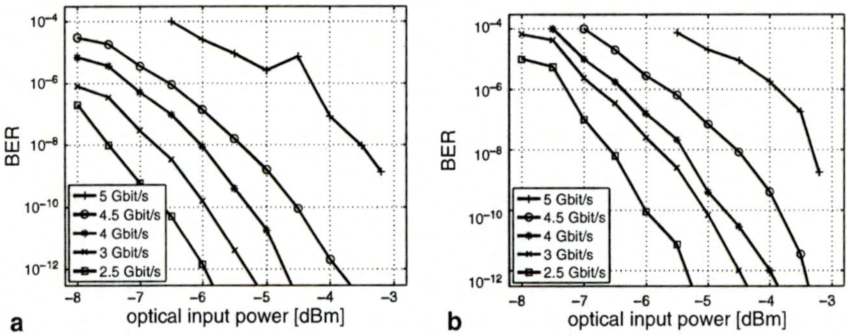

Fig. 7.8 Measured BER as a function of the optical input power for different bit rates of the 4.5 Gbit/s optical receiver with an integrated photodiode in 130 nm CMOS. **a** BER performance of chip 1; the sensitivity of this chip at a BER of 10^{-12} is −5.9 dBm at 2.5 Gbit/s, −5.3 dBm at 3 Gbit/s, −4.7 dBm at 4 Gbit/s and −3.8 dBm at 4.5 Gbit/s; the required BER level is not reached for a bit rate of 5 Gbit/s. **b** BER performance of chip 2; the sensitivity of this chip at a BER of 10^{-12} is −5.3 dBm at 2.5 Gbit/s, −4.5 dBm at 3 Gbit/s, −4 dBm at 4 Gbit/s and −3.4 dBm at 4.5 Gbit/s; the required BER level is not reached for a bit rate of 5 Gbit/s

The BER performance[2] of the developed optical receiver is shown in Fig. 7.8a for chip 1 and in Fig. 7.8b for chip 2. As expected, the BER drops if the bit rate is decreased which is due to the finite bandwidth of the receiver which closes the eye for higher bit rates. The BER also drops if the optical input power is increased. This is also as expected because noise ruins the performance for the lowest optical input powers. The smaller eye opening of chip 2 is translated into a slightly higher BER. The sensitivity is defined here as the optical input power that is required to obtain a BER below 10^{-12}. As can be deduced from Fig. 7.8a, the sensitivity of chip 1 is −5.9 dBm at 2.5 Gbit/s, −5.3 dBm at 3 Gbit/s, −4.7 dBm at 4 Gbit/s and −3.8 dBm at 4.5 Gbit/s. The required BER level is not reached if the bit rate is 5 Gbit/s. Therefore, this chip is labeled as a 4.5 Gbit/s optical receiver. Equivalently, from Fig. 7.8b it can be derived that the sensitivity of chip 2 is −5.3 dBm at 2.5 Gbit/s, −4.5 dBm at 3 Gbit/s, −4 dBm at 4 Gbit/s and −3.4 dBm at 4.5 Gbit/s, which are all slightly higher than those of chip 1. Again, the BER threshold of 10^{-12} is not obtained for a bit rate of 5 Gbit/s.

In Table 7.1 the performance of the realized optical receiver is compared to the optical receivers that have been published recently in the open literature. Remember that the proposed FOM in (7.1) is only defined for differential implementations with an integrated photodetector in a standard CMOS technology. It can be seen in this

[2] In order to quantify exactly the BER, the chip should be tested for an infinitely long time because the errors that can occur have a random nature. Consequently, the BER can only be determined up to a certain confidence level. From the formulas that are presented in [Max10] it can be estimated that a measurement time of at least 4605 s, or almost 77 min, is required to measure a BER of 10^{-12} with a confidence level of 99% if the bit rate equals 1 Gbit/s. Obviously, if the bit rate increases the required measurement time decreases accordingly for the same BER and confidence level.

Table 7.1 Comparison of the realized optical receiver in both versions to recently published optical receivers that are also integrated in a CMOS process and make use of an integrated photodiode and a differential circuit topology

	[Che07a]	[Che07b]	[Kao09]	[Lee10]	chip 1	chip 2
bit rate [Gbit/s]	2.5	3.125	5	8.5	4.5	4.5
BER	10^{-12}	10^{-12}	10^{-12}	10^{-12}	10^{-12}	10^{-12}
sensitivity [dBm]	−4.5	−4.2	−3	−3.2	−3.8	−3.4
gain [dBΩ]	110	110	124	120	105	105
PD diameter [μm]	50	65	75	70	60	60
power [mW]	108	145	183	47	138	74.2
L_{min} [nm]	180	180	180	130	130	130
FOM	10.6	15.6	21.2	242	33.3	55.3

table that chip 1 performs very well compared to other known receivers, as can be concluded from the calculated FOM. However, due to its significantly lower power consumption and the only marginally worse sensitivity, chip 2 has an even higher FOM.

7.3 A 5.5-Gbit/s Optical Receiver with a Speed-Enhanced Photodiode in 130-nm CMOS

The objective of this implementation is to demonstrate the working principle of the speed-enhanced photodiode which has been proposed in Sect. 3.2.2. Like the implementation in the previous section, this receiver is intended to be used as the analog front-end of a high-speed optical communication system. A MMF with a core diameter of 50 μm is used to guide modulated light with a wavelength of 850 nm from the sender to the receiver. As before, a PRBS with a length of $2^7 - 1$ is used to evaluate the performance of this chip.

7.3.1 *The Chip*

The architecture of this implementation is shown in Fig. 7.9. From a topology point of view, the only difference between this architecture and the architecture of the previous implementation is the photodiode. However, the design of the TIA has been adapted to function optimally in cooperation with this new photodiode. The rest of the building blocks have the same circuit implementation as before. Therefore, only the photodiode and the TIA are discussed below. All building blocks are again integrated on the same chip in a 130 nm CMOS technology.

The Photodiode To comply with the core size of a MMF, the diameter of the speed-enhanced photodiode equals 60 μm. An n-well/p-sub junction has again been selected

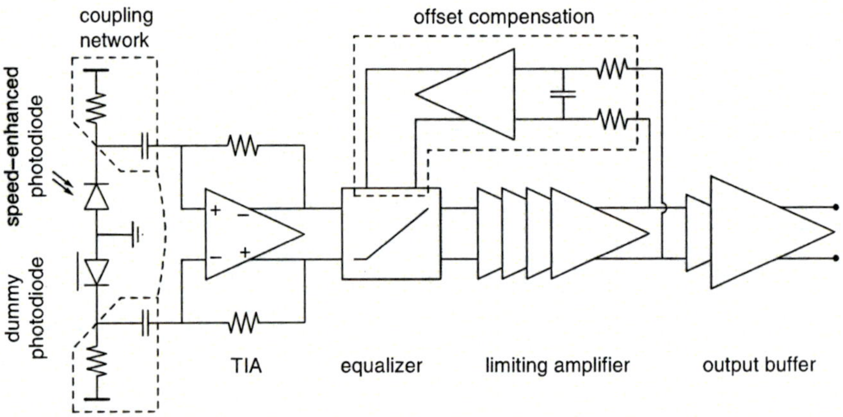

Fig. 7.9 Architecture of the 5.5 Gbit/s optical receiver with a speed-enhanced photodiode in 130 nm CMOS

due to its superior performance compared to other available junction types in a CMOS process. Two different layouts of a speed-enhanced photodiode are shown in Fig. 7.10.

The layout in Fig. 7.10a has a structure which is symmetrical around the photodiode center where a substrate contact is provided. While all other substrate contacts

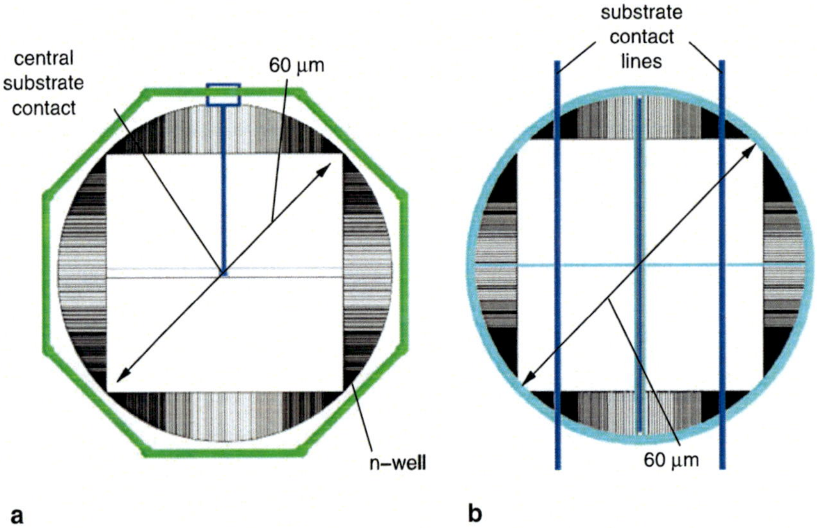

Fig. 7.10 Layout of the speed-enhanced n-well/p-sub photodiode. **a** Speed-enhanced photodiode with a single central substrate contact. **b** Speed-enhanced photodiode with two lines of substrate contacts

around the circular photodiode are kept at the ground potential of 0 V, this central substrate contact is connected to V_{field} in order to create an electric field in the substrate underneath the n-well of the photodiode junction. As explained in Sect. 3.2.2, this is done to remove the slowly and speed-limiting diffusing carriers that are generated deep in the substrate when light impinges on the photodiode. In order to obtain a fairly uniform electric field in the substrate, the substrate potential should vary from V_{field} in the center of the photodiode to 0 V at the photodiode border in a more or less linear way. However, this is not the case with the photodiode layout in Fig. 7.10a because the major portion of the potential drop is realized near the center of the structure. This can most easily be understood by considering the substrate resistance between the central substrate contact and the outer border of the photodiode. Due to its large physical size near the photodiode border and its small size near the center, this resistance has a small value near the border while being very large near the center. The consequence is that the electric field has a large magnitude near the center while it only has a very small magnitude further away from the center. As most of the diffusing carriers are generated relatively far from the center, this layout is not a very efficient one.

The second layout of a speed-enhanced photodiode is shown in Fig. 7.10b. In contrast to applying the substrate bias voltage V_{field} only to a central substrate contact, it is applied to two substrate contact lines one of which is located in the middle of the left part of the photodiode and the other is placed in the middle of the right part. A line of substrate contacts which is connected to the ground potential is placed in the middle and at both sides of the photodiode, all parallel to each other. The problem of the first layout is circumvented by this second layout as the voltage drop has a completely linear behavior between two neighboring substrate contact lines. Consequently, the electric field generated in the substrate is independent of the location so that every carrier experiences the same acceleration. Therefore, this layout has been selected.

The simulated parasitic photodiode capacitance is 860 fF if the reverse voltage over the junction is 1.2 V. This is smaller than what has been achieved with the differential photodiode in the previous section which is due to the fact that a differential photodiode has a larger perimeter for the same area and consequently a larger sidewall capacitance. The photodiode is again connected to the TIA by means of a coupling network. The resistor in this network has a resistance of 2.5 kΩ which is 8 times smaller than in the previously discussed receiver. This has been done to limit the voltage drop over this resistor[3] which is important because the DC current of this speed-enhanced photodiode is expected to be much larger than that of a differential photodiode. The capacitor in the coupling network has a capacitance of 14.5 pF. The cut-off frequency which is caused by this coupling network is then located at 4.04 MHz which is low enough for a PRBS with a length of $2^7 - 1$ and a bit rate of at least 513 Mbit/s, as can be calculated with (2.14).

[3] A voltage drop across this resistor directly affects the reverse voltage across the photodiode. Ultimately the photodiode can even become forward biased if the DC photocurrent is very large. This should obviously be prevented in order to maintain proper operation.

A dummy photodiode has been placed at the other input of the differential TIA so as to obtain a stable and symmetrical system. This dummy photodiode is covered with a metal layer to prevent scattered or diffused light from generating a photocurrent in it with the same phase as the current from the real photodiode. This would jeopardize the sensitivity of the receiver and should consequently be prevented.

The TIA The TIA has the same topology as the one that is used in the previous implementation. However, the first stage of the voltage amplifier in the TIA has been redesigned in order to take into account the smaller photodiode capacitance. While in the TIA that has been presented in Sect. 4.4.1 and used in the 4.5 Gbit/s optical receiver of the previous section the gate-source capacitance of the input transistor of the first amplifier stage has a value of 320 fF it has been halved here to 160 fF. The same size reduction by a factor of 2 has been applied on the subsequent three amplifier stages. This has been done to increase the power efficiency of the previous design even more without sacrificing the noise performance too much.

The simulated transimpedance of this new TIA is 9.06 kΩ, 1.83 times higher than in the other design, and the bandwidth is 2.83 GHz if the speed-enhanced photodiode of Fig. 7.10b and its dummy are connected at the inputs which is about the same as with the other TIA. Although the gain of the voltage amplifier did not change due to its size reduction, its smaller input capacitance together with the smaller photodiode capacitance are responsible for the better trade-off between transimpedance and bandwidth. The input noise current is now 673 nA_{RMS} while the current consumption is only 15.7 mA. Compared to the second TIA design in Sect. 4.4.1, the noise current is approximately the same whereas the current consumption has been reduced by a factor of 2. It can thus be concluded that the power efficiency is much better in this design.

7.3.2 Measurement Setup

The measurement setup is the same as with the receiver of the previous section. Its schematical diagram can be seen in Fig. 7.3.

7.3.3 Measurement Results

The design has been integrated in the same 130 nm CMOS process as the previously discussed optical receiver. The chip photograph can be seen in Fig. 7.11. A collection of four structures can be noticed in the upper part of the photograph. The structure on the left hand side is the actual speed-enhanced photodiode of which the simplified layout is depicted in Fig. 7.10b. It contains two lines of substrate contacts that are biased with a voltage which is different from the ground potential in order to generate an electric field in the substrate. The second structure is another speed-enhanced photodiode with only a single line of substrate contacts to generate the electric field.

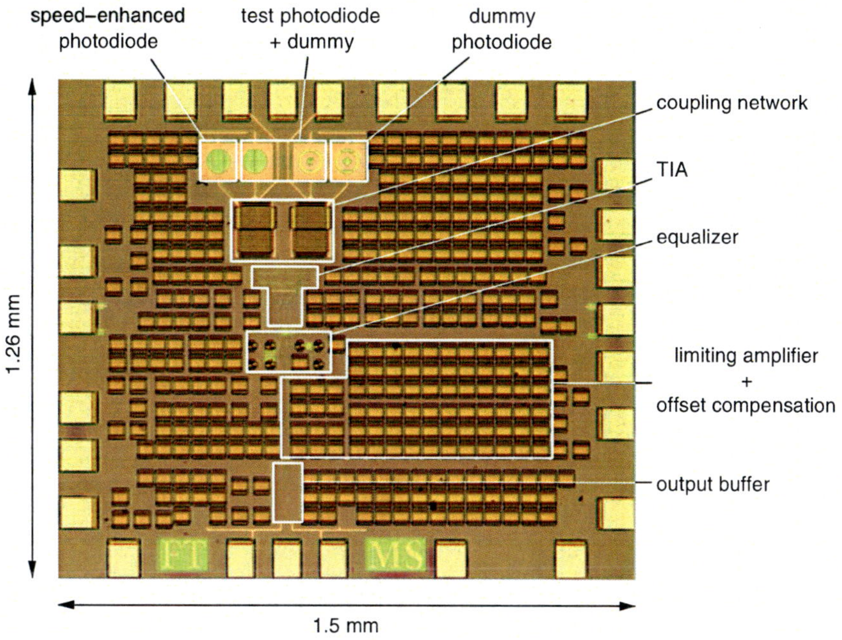

Fig. 7.11 Chip photograph of the 5.5 Gbit/s optical receiver with a speed-enhanced photodiode in 130 nm CMOS

The third structure is the dummy of the photodiode with one line of substrate contacts. It has the same layout as the actual photodiode but it is covered with metal so that it does not generate a photocurrent. The fourth structure is the dummy of the speed-enhanced photodiode with two substrate contact lines. Differential loading of the TIA is guaranteed by using the same layout for the actual photodiode and its dummy. Only the measurements of the left photodiode and its corresponding dummy are discussed below because its performance is comparable to the photodiode with only a single line of substrate contacts while it requires only half of the biasing voltage to generate the same electric field, which also follows from (3.35). The chip area is 1.89 mm^2 including the biasing circuits, decoupling capacitors and bondpads. The chip without the output buffer consumes a current of 48.8 mA at a supply voltage of 1.2 V, resulting in a power consumption of 58.6 mW.

The BER performance of this receiver is summarized in Fig. 7.12. If the electric field in the substrate underneath the speed-enhanced photodiode is not activated, it can be derived from Fig. 7.12a that the sensitivity of this chip at a BER threshold of 10^{-12} is −7.8 dBm at 2.5 Gbit/s, −7.3 dBm at 3 Gbit/s, −5.7 dBm at 4 Gbit/s, −5.4 dBm at 4.5 Gbit/s, −4.3 dBm at 5 Gbit/s and −2.7 dBm at 5.5 Gbit/s. Although a bit rate of 6 Gbit/s can also be received, the required BER threshold is not reached. Applying a voltage difference of 1 V between the two substrate contact lines and the ground potential improves the BER performance of the receiver considerably, as can be seen in Fig. 7.12b. The sensitivity is now −7.8 dBm at 2.5 Gbit/s, −7.3 dBm at 3 Gbit/s, −5.7 dBm at 4 Gbit/s, −5.7 dBm at 4.5 Gbit/s, −4.9 dBm

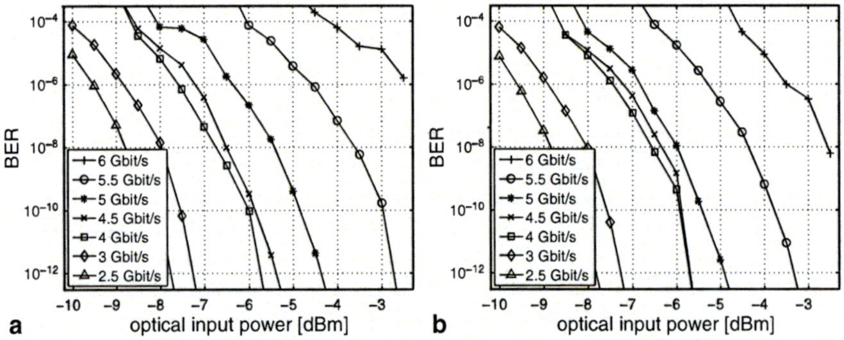

Fig. 7.12 Measured BER as a function of the optical input power for different bit rates of the 5.5 Gbit/s optical receiver with a speed-enhanced photodiode in 130 nm CMOS. **a** BER performance of the receiver without an electric field in the substrate of the speed-enhanced photodiode; the sensitivity of this chip at a BER of 10^{-12} is −7.8 dBm at 2.5 Gbit/s, −7.3 dBm at 3 Gbit/s, −5.7 dBm at 4 Gbit/s, −5.4 dBm at 4.5 Gbit/s, −4.3 dBm at 5 Gbit/s and −2.7 dBm at 5.5 Gbit/s; the required BER level is not reached for a bit rate of 6 Gbit/s. **b** BER performance of the receiver with an electric field in the substrate of the speed-enhanced photodiode; the sensitivity of this chip at a BER of 10^{-12} is −7.8 dBm at 2.5 Gbit/s, −7.3 dBm at 3 Gbit/s, −5.7 dBm at 4 Gbit/s, −5.7 dBm at 4.5 Gbit/s, −4.9 dBm at 5 Gbit/s and −3.4 dBm at 5.5 Gbit/s; the required BER level is not reached for a bit rate of 6 Gbit/s

at 5 Gbit/s and −3.4 dBm at 5.5 Gbit/s. A 6 Gbit/s signal can again be received although not up to the required BER level. By comparing Fig. 7.12a and 7.12b, it can be concluded that the performance at the highest bit rates is improved by activating the electric field underneath the speed-enhanced photodiode. This is as expected because the performance at the lower bit rates is determined mainly by the noise of the receiver while the performance at higher bit rates is limited by the finite bandwidth of the photodiode which is improved by the electric field.

The performance of this receiver is summarized in Table 7.2 where it is also compared to the receivers that were presented in the previous section and to some currently published optical receivers with an integrated photodiode. The FOM of this

Table 7.2 Comparison of the realized optical receiver to recently published optical receivers that are also integrated in a CMOS process and make use of an integrated photodiode and a differential circuit topology; chip 1 and chip 2 are the two versions of the receiver that were presented in Sect. 7.2

	[Che07a]	[Che07b]	[Kao09]	[Lee10]	chip 1	chip 2	this chip
bit rate [Gbit/s]	2.5	3.125	5	8.5	4.5	4.5	5.5
BER	10^{-12}	10^{-12}	10^{-12}	10^{-12}	10^{-12}	10^{-12}	10^{-12}
sensitivity [dBm]	−4.5	−4.2	−3	−3.2	−3.8	−3.4	−3.4
gain [dBΩ]	110	110	124	120	105	105	124
PD diameter [μm]	50	65	75	70	60	60	60
power [mW]	108	145	183	47	138	74.2	58.6
L_{min} [nm]	180	180	180	130	130	130	130
FOM	10.6	15.6	21.2	242	33.3	55.3	101

receiver is even higher than the FOM of both 4.5 Gbit/s receivers of the previous section which is mainly due to the higher attainable bit rate and the lower power consumption.

7.4 A POF Receiver with a 1-mm Diameter Integrated Photodiode in 180-nm CMOS

The optical receiver that is presented in this section is intended to be used together with a POF with a core diameter of 1 mm. Instead of the 850 nm light which is used in a MMF, light with a wavelength of 660 nm is generally applied with such a plastic fiber. A PRBS with a length of $2^7 - 1$ is employed to evaluate the transient behavior of the chip. The main objective of this receiver is to combine high speed and low power consumption under the limitations which are imposed by the large photodiode and its correspondingly large parasitic capacitance.

7.4.1 *The Chip*

The architecture of this receiver is visualized in Fig. 7.13. It consists of a photodiode, a TIA with DC current compensation, an equalizer and an output buffer. In contrast with the previous implementations, no post amplifier is provided in this receiver chain. All building blocks are integrated on the same chip in a 180 nm CMOS process.

The Photodiode The n-well/p-sub integrated photodiode has a diameter of 1 mm, which is just as large as the diameter of a typical POF. A single-ended photodiode

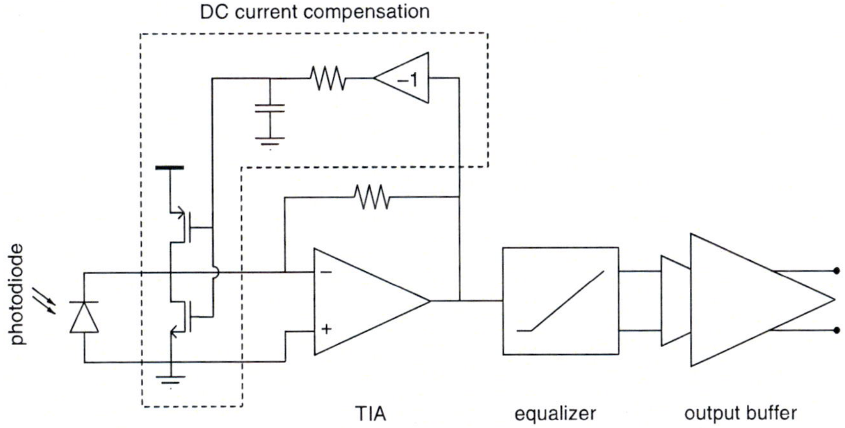

Fig. 7.13 Architecture of the POF receiver with a 1 mm diameter integrated photodiode in 180 nm CMOS

architecture has been selected for several reasons. Firstly, a differential photodiode has a much larger parasitic capacitance because it has a large perimeter for a certain area. Secondly, a differential photodiode has an intrinsically lower responsivity because half of the light is reflected. Thirdly, the intended bit rates with this receiver are about 10 times lower than those that have been achieved with the previously discussed implementations. Finally, as a wavelength of 660 nm is used instead of 850 nm, the intrinsic bandwidth of the photodiode is higher due to the higher absorption coefficient of silicon at this shorter wavelength. According to [Rad03], the bandwidth of an integrated n-well/p-sub photodiode equals approximately 70 MHz if 660 nm light is used.

The simulated parasitic capacitance of this structure is 63.6 pF for a reverse voltage of 0.56 V which is the reverse voltage at which it is biased during operation. Obviously, this huge value is the result of the large photodiode area. However, the technology limitations regarding the implementation of photodiodes may not be underestimated. As a reference, it can be compared with the parasitic capacitance of 2.2 pF that has been obtained in [För06] where a photodiode with a diameter of 0.5 mm has been integrated in a 0.6 μm BiCMOS process.

The TIA The TIA that is used in this receiver is the one that has been introduced in Sect. 4.4.2. A DC current compensation circuit is provided which cancels out the DC current of the photodiode and thereby removes the offset of the photocurrent. Signal frequencies that can pass through the low-pass filter of this compensation circuit are exposed to a transimpedance which is determined by the transconductance of its nMOS and pMOS which are in parallel with the photodiode. On the contrary, higher signal frequencies experience a transimpedance which is determined only by the feedback resistance of the TIA. Consequently, the transimpedance of such a TIA has a high-pass characteristic of which the lower pole frequency determines the lowest signal frequency.

As discussed in Sect. 4.4.2, the simulated transimpedance of this TIA is 56.9 kΩ which is maintained over a bandwidth of 139 MHz if the photodiode from above is attached at its input. The transimpedance at low frequencies is only 1.49 kΩ which is low enough so that a large DC photocurrent cannot cause the TIA to shift from its ideal biasing point. The cut-off frequency between both transimpedance values is located at 8.76 kHz. This is low enough for the $2^7 - 1$ PRBS at the intended bit rates. The overdrive voltage of the amplifying transistors in the TIA has been chosen equal to 50 mV in order to obtain good power efficiency[4]. The input noise current has a simulated value of 137 nA_{RMS} for a current consumption of 13.7 mA. As the supply voltage equals 1.12 V for the desired operating point, the power consumption equals 15.3 mW. By increasing the supply voltage, the current consumption is increased accordingly and the bandwidth of the TIA is extended significantly, as shown below.

The Equalizer The equalizer in this implementation is the design that has been introduced in Sect. 5.3.2. Instead of making use of the principle of frequency dependent source degeneration, which has been applied in the equalizer of the previously

[4] This overdrive voltage is regulated by means of the supply voltage which has a direct relationship with it, as can be seen in the circuit diagram in Fig. 4.37.

discussed optical receivers, this equalizer has three active inductors in parallel with each other to generate the desired tuning characteristic. It has been shown in Sect. 5.3.2 that this equalizer is able to compensate the roll-off of an n-well/p-sub photodiode up to approximately 500 MHz. The resistance of the linear pMOS transistors in the active inductors can be altered by changing their gate voltages during operation. This equalizer has a power consumption of 297 μW if the supply voltage equals 1.12 V, which is the same as the supply voltage of the TIA.

The Output Buffer By contrast with the TIA and the equalizer which are single-ended circuits, the output buffer has a differential implementation which has been done to limit the high frequency current spikes through the supply and ground bondwires. It is composed of a cascade of four stages with a tapered design so that it presents a low load capacitance to the equalizer while delivering enough current to the 50 Ω input impedance of the measurement equipment.

7.4.2 *Measurement Setup*

The measurement setup for this chip resembles the setup that is shown in Fig. 7.3. However, the laser diode has been replaced by a LED with a center wavelength at 660 nm while the fiber has been replaced by a POF. No lens has been attached at the end of this fiber to focus the light on the integrated photodiode. On the contrary, the fiber has been cut and stripped with simple tools and positioned directly above the photodiode. Due to the large diameter of the fiber and the photodiode this was relatively easy to achieve.

As can be seen in Fig. 7.14, the chip has been mounted on a small PCB which provides two 50 Ω transmission lines for the differential output signal. Furthermore,

Fig. 7.14 Photograph of the POF receiver with a 1 mm diameter integrated photodiode in 180 nm CMOS mounted on a small PCB for test purposes

it provides several metal tracks which route the biasing voltages and currents towards the chip as well as some decoupling capacitors. This small PCB has been mounted on a larger PCB taking care of the generation of all required voltages and currents.

The output signal of the chip is visualized by means of an oscilloscope instead of being analyzed by the analyzer module of the BER tester of which the generator module generates the PRBS. This has been done because the analyzer module is unable to operate at the bit rates that can be processed by the presented POF receiver.

7.4.3 Measurement Results

The chip photograph of this 180 nm CMOS design is shown in Fig. 7.15. The total chip area including bondpads, biasing circuits and decoupling capacitors amounts

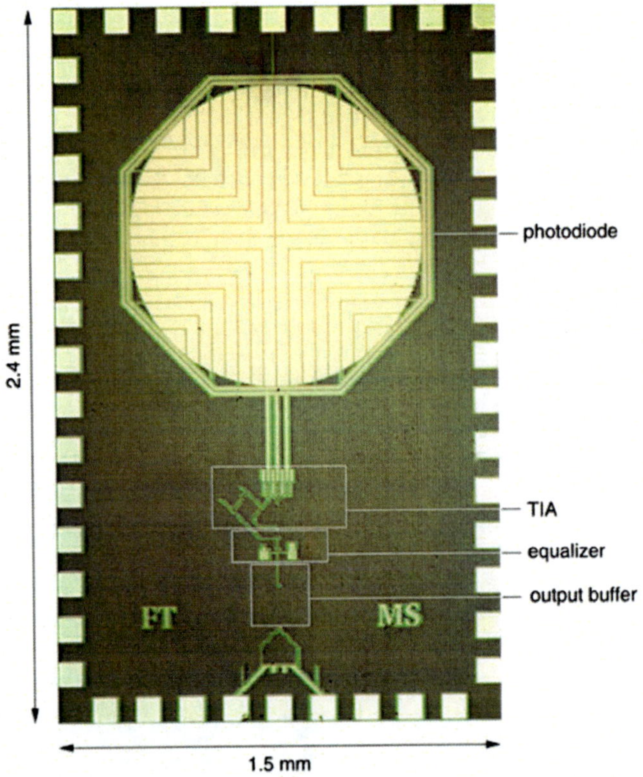

Fig. 7.15 Chip photograph of the POF receiver with a 1 mm diameter integrated photodiode in 180 nm CMOS

to 3.6 mm^2. Almost half of this area is consumed by the photodiode, as can be seen in the chip photograph. Only a small portion of it is required for the active circuits which have been placed far from the photodiode. This has been done in order not to jeopardize their performance due to processing issues that could appear which are caused by the large area that is kept free of any metal layer. The chip consumes 39.3 mW of power if the supply voltage equals 1.12 V.

The transient performance of this receiver has not been assessed by means of its BER performance. On the contrary, its output signal has been visualized with an oscilloscope on which the eye diagram has been analyzed. The most important **performance figure regarding such an eye diagram is its horizontal eye opening,** which is affected by the amount of jitter on the signal. The measured peak-to-peak jitter, normalized to the bit period, of this POF receiver is summarized in Fig. 7.16a for the nominal supply voltage of 1.12 V and in Fig. 7.16b if the supply voltage is

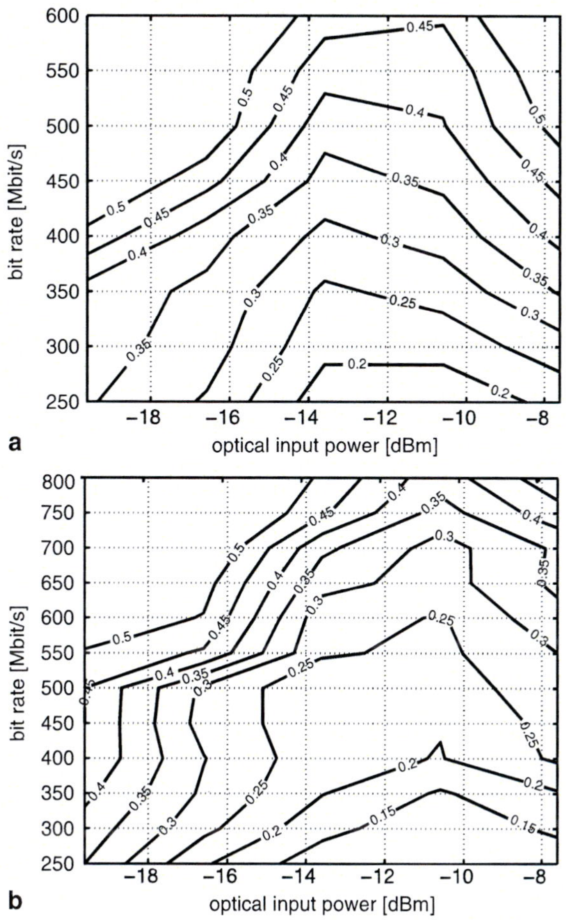

Fig. 7.16 Contour plots of the measured peak-to-peak jitter normalized to the bit period as a function of the optical input power and the bit rate for two different supply voltages. **a** Normalized peak-to-peak jitter if the supply voltage is 1.12 V; a bit rate up to 300 Mbit/s can be received properly for an optical input power between −14.5 and −9 dBm. **b** Normalized peak-to-peak jitter if the supply voltage is 1.42 V; a bit rate up to 500 Mbit/s can be received properly for an optical input power between −15 and −9.1 dBm

increased to 1.42 V, at which the transistors are biased with an overdrive voltage of 200 mV instead of the 50 mV at the nominal supply voltage. From both figures it can be concluded that the amount of normalized jitter increases at higher bit rates. Moreover, it can be noticed that the normalized jitter varies with the optical input power. If a horizontal eye opening of at least 75% of the bit interval is required, it can be derived from Fig. 7.16a that at a supply voltage of 1.12 V a bit rate of 300 Mbit/s can be received over a reasonable range of optical input powers, namely between −14.5 and −9 dBm. For the same horizontal eye opening, it can be derived from Fig. 7.16b that at a supply voltage of 1.42 V a bit rate of 500 Mbit/s can be processed correctly if the optical input power is between −15 and −9.1 dBm.

Not a single CMOS receiver with an integrated 1 mm diameter photodiode has been found in the open literature until now which can be used to put into perspective the performance of this POF receiver. However, a POF receiver is proposed in [För06] which is integrated in a 0.6 μm BiCMOS process. This technology makes it possible to construct a PIN photodiode with a very wide depletion region. Consequently, the integrated photodiode with a diameter of 0.5 mm has a parasitic capacitance of only 2.2 pF at a reverse voltage of 2.5 V. The receiver has a bandwidth of 220 MHz for a power consumption of 27.3 mW. An ARC has been applied to obtain a sensitivity below −16.1 dBm. It can consequently be concluded that the POF receiver with a 1 mm diameter photodiode in 180 nm CMOS has a comparable bandwidth, power consumption and sensitivity while being integrated in a standard CMOS process, having a 4 times larger photodiode and not using an ARC.

7.5 A Low Power and Area Efficient Limiting Amplifier in 90-nm CMOS

This chip integrates only a single building block of an optical receiver, namely a limiting amplifier. It is intended to be used in a high-speed communication system of which the transmitted signals can be modelled with a $2^{31} - 1$ PRBS. The main objective of this chip is to combine high speed, small chip area and low power consumption.

7.5.1 The Chip

The chip architecture is shown in Fig. 7.17. The input buffer takes care of the 50 Ω input termination and subtracts the measured offset voltage from the input voltage. The limiting amplifier boosts the signal swing and its offset voltage is compensated out by means of an active offset compensation circuit. The output buffer has been added to be able to deliver enough current to the 50 Ω input impedance of the measurement equipment. All building blocks are integrated on a single chip in 90 nm CMOS.

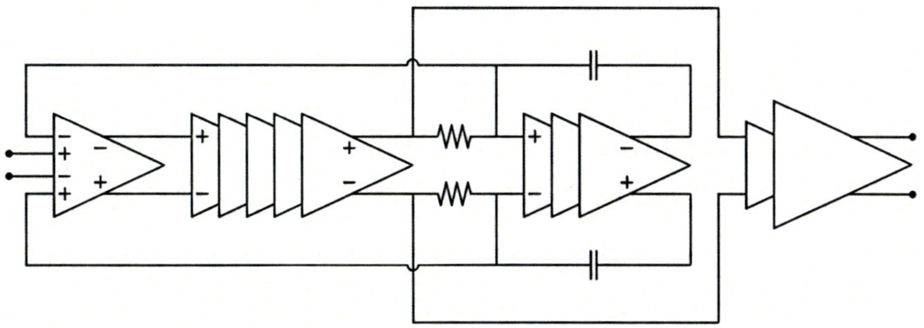

Fig. 7.17 Architecture of the low power and area efficient limiting amplifier in 90 nm CMOS

The Input Buffer, Limiting Amplifier and Offset Compensation Circuit The input buffer, limiting amplifier and offset compensation circuit are those that have been introduced in Sect. 6.5.2. The input buffer is a differential difference amplifier, where the voltage gain from each of the inputs to the output is approximately 0 dB. A 100 Ω resistor has been placed differentially between the inputs of the chip so as to take care of the input termination. The amplifier core is composed of a cascade of five first-order stages, as can be seen in the circuit diagram in Fig. 6.19. As discussed in Sect. 6.3.1, this is optimal regarding both the required GBW per stage and the total power consumption for a voltage gain around 40 dB. Negative Miller capacitors have been added between the stages to increase the bandwidth of the cascade. The low-pass filter of the offset compensation circuit is composed of a resistor and an actively boosted capacitor so as to increase the effective time constant by the gain of the boosting amplifier. This has been done in order to reduce the area which is required to integrate the passive components of the offset compensation circuit.

The simulated voltage gain of this amplifier is 37.5 dB in the frequency band of interest while the voltage gain at DC is 0 dB due to the offset compensation feedback loop. A lower cut-off frequency is located at 925 kHz which is, according to (6.39), not low enough to amplify all frequency components of a PRBS with a length of $2^{31} - 1$. It requires that the lower corner frequency is located below 1 Hz if the bit rate is 1 Gbit/s. However, such a low corner frequency would jeopardize the settling time of the amplifier. Therefore, a value of 925 kHz has been chosen as a trade-off between baseline wander and settling time. The bandwidth of the limiting amplifier after parasitic extraction is 4.15 GHz. The input noise voltage is only 285 μV_{RMS} for a total power consumption of 12.7 mW at a supply voltage of 1 V.

The Output Buffer The output buffer consists of seven cascaded differential stages that each have a voltage gain of around 0 dB. Seven stages have been used so as to minimize the load capacitance of the limiting amplifier while being able to steer enough current through the 50 Ω input impedance of the measurement equipment. The rate at which the buffer stages have been scaled up towards the output have been kept small thanks to the large number of stages. Consequently, the small-signal

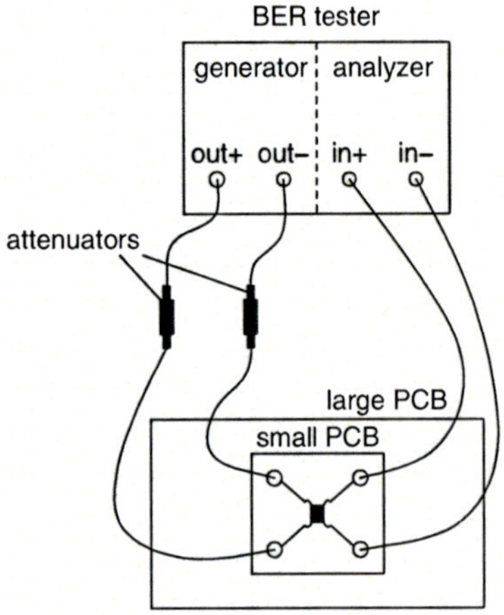

Fig. 7.18 Schematical representation of the transient measurement setup of the integrated limiting amplifier

bandwidth of the output buffer has a simulated value of 3.65 GHz after parasitic extraction.

7.5.2 Measurement Setup

The transient measurements on the chip have been carried out using the setup that is shown in Fig. 7.18. The generator module of the BER tester generates a PRBS with the appropriate length. Its output signal is attenuated in order to be able to test the chip down to its noise level. The outputs of the attenuators are connected to the small PCB on which 50 Ω transmission lines are provided to route the input signal towards the chip which is mounted on it. In its turn, this small PCB is mounted on a large PCB which provides all necessary voltages and currents to bias the chip. The outputs of the chip are connected to the inputs of the analyzer module of the BER tester. The latter compares the generated and analyzed bit streams and counts the number of errors to obtain the BER of the chip.

The schematical representation of the measurement setup to assess the small-signal behavior of the chip for frequencies below 800 MHz is shown in Fig. 7.19a. The single-ended output of a vector network analyzer is transformed into a differential signal by means of a balun[5]. An attenuator is placed in between in order to make sure that the limiting amplifier is operated in its linear region. The output signals of the chip are combined by means of another balun of which the single-ended output is connected to the input of the vector network analyzer. As the applied baluns can only be operated between 100 kHz and 800 MHz, the setup that is shown in Fig. 7.19b is applied to characterize the chip at higher frequencies. In this case, the output of the

[5] A balun, or balanced to unbalanced transformer, is a type of transformer that can be used to convert a differential signal into a single-ended signal and vice versa.

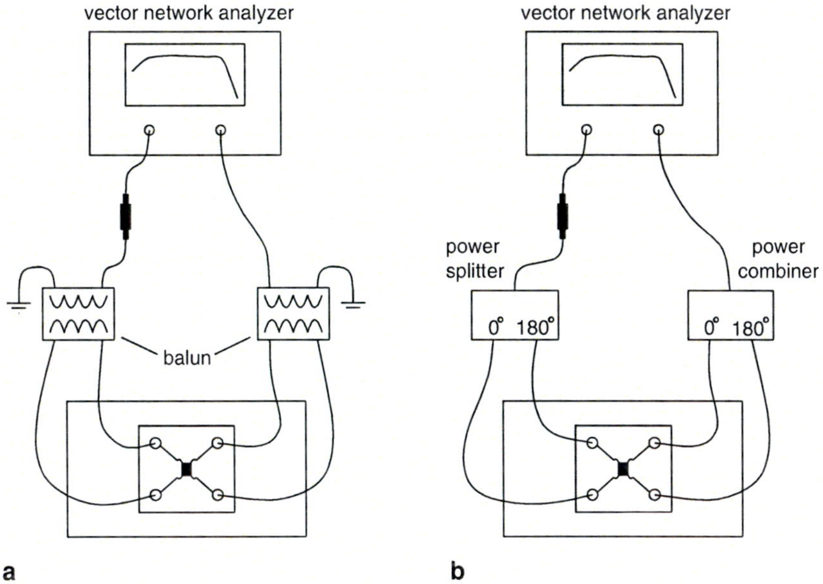

Fig. 7.19 Schematical representation of the small-signal measurement setup of the integrated limiting amplifier. **a** Setup to measure the small-signal behavior up to 800 MHz. **b** Setup to measure the small-signal behavior above 800 MHz

network analyzer is converted to a differential signal by means of a power splitter which has an operating frequency range which starts around 1 GHz. The outputs of the chip are combined by means of a power combiner. In order to characterize the performance of the chip alone, the parasitics of the two measurement setups in Fig. 7.19 are de-embedded down to the chip inputs and outputs.

7.5.3 Measurement Results

The chip photograph of the limiting amplifier which has been integrated in a 90 nm CMOS process is shown in Fig. 7.20. The chip has a total area of 0.46 mm^2 including the output buffer, biasing circuits and bondpads. The area of the limiting amplifier alone is only 0.12 mm^2. The amplifier and its biasing circuits have a power consumption of 14.7 mW at a supply voltage of 1 V, the nominal supply voltage of the applied technology.

The BER measurements of the chip are summarized in Fig. 7.21 when a PRBS with a length of $2^{31}-1$ is applied at its input. The BER drops if the input signal swing is increased, regardless of the bit rate. The BER also drops if the bit rate is decreased from 5 to 4 Gbit/s. However, if the bit rate is further decreased to 3 Gbit/s the BER does not drop anymore. Consequently, it can be concluded that the bandwidth of

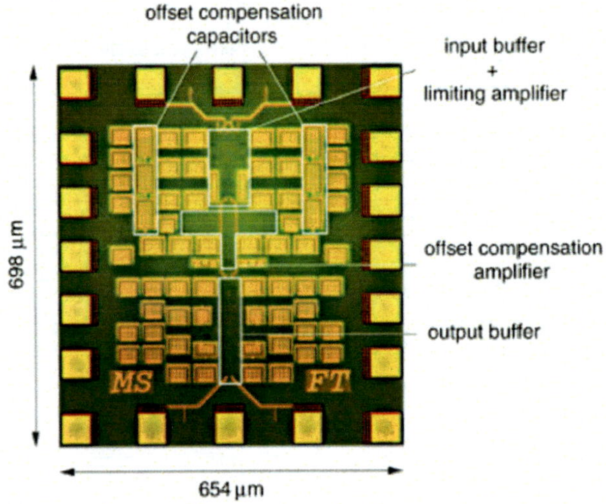

Fig. 7.20 Chip photograph of the low power and area efficient limiting amplifier in 90 nm CMOS

Fig. 7.21 Measured BER as a function of the input signal amplitude for three different bit rates

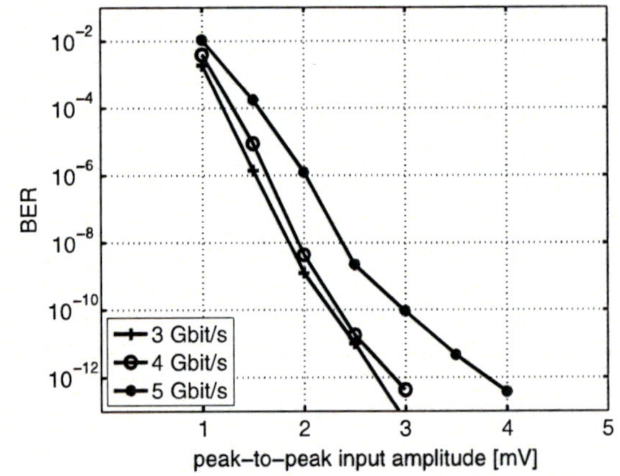

the chip is high enough in order not to introduce any noticeable DDJ at a bit rate of 3 and 4 Gbit/s. The sensitivity at a BER threshold of 10^{-12} is 2.75 mV at 3 Gbit/s, 2.9 mV at 4 Gbit/s and 3.75 mV at 5 Gbit/s. The BER stays below 10^{-12} if the input signal swing is increased above these values up to a swing of 1 V, which is the supply voltage. As a result, the useful dynamic range of this limiting amplifier is 51.2 dB, 50.8 dB and 48.5 dB for the three measured bit rates.

The magnitude of the measured small-signal transfer function of the chip is shown in Fig. 7.22. The response of the complete measurement setup, including the small PCB on which the chip has been mounted, has been calibrated out by means of a reference measurement. Therefore, it can be concluded from this figure that the small-signal gain of the cascade of the limiting amplifier and the output buffer equals

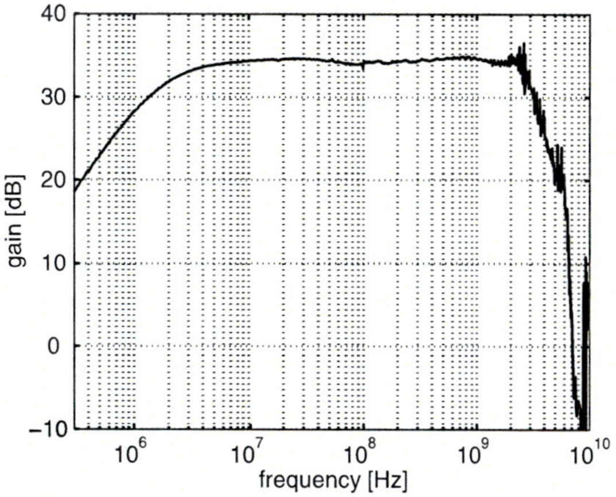

Fig. 7.22 Measured small-signal gain of the limiting amplifier, offset compensation circuit and output buffer

Table 7.3 Comparison of this chip to recently published CMOS implementations; the numbers in italics have been calculated with the knowledge that for a limiting amplifier a bandwidth of B Hz is required for a bit rate of B bit/s

	[Lia06]	[Her05]	[Nog06]	[Gal03]	[Cra06]	[San03]	[Hua07]	this chip
gain [dB]	21	27	26	50	42	40	42	35
bandwidth [GHz]	*10*	*3.5*	*10*	9.4	*3.125*	*10*	9	4.15
bit rate [Gbit/s]	10	3.5	10	*9.4*	3.125	10	*9*	*4.15*
sensitivity [mV]	18	10	40	4.6	2.5	10	10	2.9
L_{min} [nm]	180	180	130	180	180	180	180	90
power [mW]	54	34	500	100	113	144	189	14.7
area [mm^2]	1.32	1.95	2.72	0.75	0.5	0.4	0.192	0.12
FOM	2833	3829	62	279160	90241	56250	101164	369519

35 dB. The lower cut-off frequency is located at 1.86 MHz whereas the bandwidth of the cascade equals 3.11 GHz, which is exactly the same as the simulated bandwidth after parasitic extraction. It can thus be concluded that the limiting amplifier on its own reaches its simulated bandwidth of 4.15 GHz.

The presented limiting amplifier is compared to some recently published implementations in Table 7.3. The FOM that is used in this table is defined as follows:

$$\text{FOM} = \frac{\text{GBW [GHz]} \cdot L_{min}^2 \text{ [nm}^2]}{\text{sensitivity [mV]} \cdot \text{power [mW]} \cdot \text{area [mm}^2]}. \tag{7.2}$$

L_{min} is squared in this FOM in order not to favor implementations which are made in a technology with a smaller minimal gate length and a correspondingly smaller area. The presented receiver has the highest FOM in Table 7.3 which is especially thanks to its very good sensitivity, its low power consumption and its small area.

7.6 Conclusion

After having treated the photodiode, the TIA, the equalizer and the post amplifier in detail in the previous chapters, four chip implementations are discussed in this chapter that make use of a combination of these building blocks.

In order to be able to assess the quality of an optical receiver objectively, a FOM for CMOS optical receivers with integrated photodiode is introduced first. This FOM requires that the implementation under consideration is integrated in a standard CMOS process, has an integrated photodiode, uses a differential circuit topology and is measured with 850 nm modulated light.

The first presented chip is a 4.5 Gbit/s optical receiver with an integrated photodiode in 130 nm CMOS. This high speed operation is realized by means of a differential photodiode and an optimized design of the differential TIA. An equalizer boosts the bandwidth of the differential photodiode even more. The sensitivity of this chip is −3.8 dBm or −3.4 dBm, depending on the design of the TIA and it has a power consumption of 138 mW or 74.2 mW, respectively.

The second chip is a 5.5 Gbit/s optical receiver with a speed-enhanced photodiode in 130 nm CMOS. A speed-enhanced photodiode is used in this design instead of the differential photodiode of the first presented chip. An equalizer is again added to increase the bandwidth of this photodiode. This receiver realizes a sensitivity of −3.4 dBm at a power consumption of only 58.6 mW.

The third optical receiver is integrated in 180 nm CMOS and is intended to be used with a POF instead of with a MMF. Consequently, its integrated photodiode has a diameter of 1 mm to comply with the core diameter of a POF. Thanks to an optimized TIA design, a bit rate up to 300 Mbit/s can be received. This POF receiver has a power consumption of 39.3 mW at a supply voltage of 1.12 V. If the supply voltage is increased to 1.42 V, the maximum bit rate is increased to 500 Mbit/s.

The last chip implementation that is presented is a low power and area efficient limiting amplifier in 90 nm CMOS. The power efficiency is realized by a deliberate sizing of the amplifier stages and the use of negative Miller capacitors. The area efficiency is obtained by boosting the time constant of the low-pass filter in the offset compensation circuit by means of a boosting amplifier. This limiting amplifier has a measured gain and bandwidth of 35 dB and 4.15 GHz, respectively. The sensitivity of this design is 2.9 mV and the chip consumes 14.7 mW of power.

Chapter 8
Conclusions

8.1 General Conclusions

In this work, the design, assessment and measurement of optical receivers with integrated photodiode in nanoscale CMOS are treated in detail. The different building blocks of an optical receiver chain are examined in the first chapters. Based on these discussions, the realized chip implementations are presented thereafter.

An introduction in the domain of optical communication is given in Chap. 1. The questions *Why do we need optical communication?* and *What are integrated optical receivers?* are answered first. Next, an overview on the relevant publications in the open literature is given, which is subdivided into three categories: non-CMOS implementations, CMOS implementations without an integrated photodiode and CMOS implementations with an integrated photodiode.

A high-level perspective on the field of optical communication is provided in Chap. 2. The communication model is introduced and the properties of random binary data are discussed. The working principle of a laser diode and an optical fiber are also explained. In the remainder of the chapter, the fundamentals of an optical receiver are presented and the three dominant trade-offs in their design are derived: bandwidth versus bit rate, noise versus bandwidth and bit error ratio versus noise.

In Chap. 3, a detailed discussion of the photodiode is given. Firstly, the working principle is explained, after which the possibilities of CMOS technologies to integrate photodiodes are assessed. Several techniques are proposed to improve the speed of a CMOS photodiode, including the speed-enhanced photodiode.

The TIA is dealt with in Chap. 4. Its important specifications are introduced and a detailed overview is given of the different TIA topologies. The shunt-shunt feedback TIA is shown to have the best overall performance. Some considerations on the design of the voltage amplifier and the feedback network of this type of TIA are treated thereafter. To conclude, three TIA implementations are shown, together with their circuit simulations.

The speed of an integrated photodiode is compensated by the equalizer which is treated in Chap. 5. Its operation principle is explained and some circuit techniques that can be applied to realize the desired tuning characteristic are introduced. Further, two circuit implementations are proposed and their simulation results are discussed.

F. Tavernier, M. Steyaert, *High-Speed Optical Receivers with Integrated Photodiode in Nanoscale CMOS*, Analog Circuits and Signal Processing 5,
DOI 10.1007/978-1-4419-9925-2_8,

The last building block of an optical receiver, namely the post amplifier, is analyzed in detail in Chap. 6. First, the arguments are presented as to why to choose a limiting amplifier over an AGC amplifier. Its main specifications are introduced, after which some circuit techniques are presented to realize a broadband amplifier. Next, several possibilities to implement an offset compensation circuit are compared. Finally, two design cases are presented.

In Chap. 7, knowledge of all the building blocks of an optical receiver is combined to discuss the realized chip implementations. The first chip is a power efficient 4.5 Gbit/s optical receiver with an integrated photodiode in 130 nm CMOS. Depending on the design of its differential TIA, it has a sensitivity of −3.8 dBm or −3.4 dBm for a power consumption of 138 mW and 74.2 mW, respectively. The second implementation discussed is a 5.5 Gbit/s optical receiver with an integrated speed-enhanced photodiode in 130 nm CMOS. At a power consumption of only 58.6 mW, it is able to obtain a sensitivity of −3.4 dBm.

The next optical receiver chip presented is a high-speed POF receiver with a 1 mm diameter integrated photodiode in 180 nm CMOS. Despite the huge parasitic capacitance of the large photodiode, it is able to properly operate at a bit rate of 300 Mbit/s or 500 Mbit/s at a supply voltage of 1.12 V and 1.42 V, respectively. The last chip implementation is a low power and area efficient limiting amplifier in 90 nm CMOS. It has a measured gain of 35 dB and a bandwidth of 4.15 GHz for a power consumption of only 14.7 mW. Thanks to an active implementation of the low-pass filter in the offset compensation circuit, the core chip area is only 0.12 mm^2.

References

[Ana06] B. Analui, D. Guckenberger, D. Kucharski, and A. Narasimha, "A Fully Integrated 20-Gb/s Optoelectronic Transceiver Implemented in a Standard 0.13-μm CMOS SOI Technology", *IEEE Journal of Solid-State Circuits*, vol. 41, no. 12, pp. 2945–2955, December 2006.

[Azu08] S. Azuma, R. Yanagimoto, S. Kamitani, M. Edamoto, K. Arata, H. Matsui, H. Akada, R. Masuda, K. Hoshino, K. Nagura, and H. Ogawa, "1.25Gbps Optical Links for Mobile Handsets", in *Proceedings of the 2008 IEEE Asian Solid-State Circuits Conference*, pp. 33–36, November 2008.

[Cha91] Z.Y. Chang and W.M.C. Sansen, *Low-Noise Wide-Band Amplifiers in Bipolar and CMOS Technologies*, Kluwer Academic Publishers, 1991.

[Che63] E.M. Cherry and D.E. Hooper, "The design of wide-band transistor feedback amplifiers", *Proceedings of the Institution of Electrical Engineers*, vol. 110, no. 2, pp. 375–389, February 1963.

[Che04] C.-C Chen, C.-M. Tsai, and L.-R. Huang, "An 1.25Gbit/s −29 dBm burst-mode optical receiver realized with 0.35μm SiGe BiCMOS process using a PIN photodiode", in *Proceedings of the 2004 IEEE Asia-Pacific Conference on Circuits and Systems*, volume 1, pp. 313–316, December 2004.

[Che05] W.-Z. Chen, Y.-L. Cheng, and D.-S. Lin, "A 1.8-V 10-Gb/s Fully Integrated CMOS Optical Receiver Analog Front-End", *IEEE Journal of Solid-State Circuits*, vol. 40, no. 6, pp. 1388–1396, June 2005.

[Che06] W.-Z. Chen and C.-H. Lu, "Design and Analysis of a 2.5-Gbps Optical Receiver Analog Front-End in a 0.35-μm Digital CMOS Technology", *IEEE Transactions on Circuits and Systems I: Regular Papers*, vol. 53, no. 5, pp. 977–983, May 2006.

[Che07a] W.-Z. Chen and S.-H. Huang, "A 2.5 Gbps CMOS Fully Integrated Optical Receicer with Lateral PIN Detector", in *Proceedings of the 2007 IEEE Custom Integrated Circuits Conference*, pp. 293–296, September 2007.

[Che07b] W.-Z. Chen, S.-H. Huang, G.-W. Wu, C.-C. Liu, Y.-T. Huang, C.-F. Chin, W.-H Chang, and Y.-Z. Juang, "A 3.125 Gbps CMOS Fully Integrated Optical Receiver with Adaptive Analog Equalizer", in *Proceedings of the 2007 IEEE Asian Solid-State Circuits Conference*, pp. 396–399, November 2007.

[Che07c] W.-Z. Chen and D.-S. Lin, "A 90-dBΩ 10-Gb/s Optical Receiver Analog Front-End in a 0.18-μm CMOS Technology", *IEEE Transactions on Very Large Scale Integration Systems*, vol. 15, no. 3, pp. 358–365, March 2007.

[Che09] W.-Z. Chen, R.-M. Gan, and S.-H. Huang, "A Single-Chip 2.5-Gb/s CMOS Burst-Mode Optical Receiver", *IEEE Transactions on Circuits and Systems I: Regular Papers*, vol. 56, no. 10, pp. 2325–2331, October 2009.

[Cou02] L.W. Couch, *Digital and Analog Communication Systems*, Prentice-Hall, Inc., Upper Saddle River, NJ 07458, sixth edition, 2002.

F. Tavernier, M. Steyaert, *High-Speed Optical Receivers with Integrated Photodiode in Nanoscale CMOS*, Analog Circuits and Signal Processing 5,
DOI 10.1007/978-1-4419-9925-2,

[Cra06] E.A. Crain and M.H. Perrott, "A 3.125 Gb/s Limit Amplifier in CMOS with 42 dB gain and 1 μs Offset Compensation", *IEEE Journal of Solid-State Circuits*, vol. 41, no. 2, pp. 443–451, February 2006.

[Csu02] S.M. Csutak, J.D. Schaub, W.E. Wu, and J.C. Campbell, "High-Speed Monolithically Integrated Silicon Optical Receiver Fabricated in 130-nm CMOS Technology", *IEEE Photonics Technology Letters*, vol. 14, no. 4, pp. 516–518, April 2002.

[ESA10] ESA, "Another world first for Artemis: a laser link with an aircraft", http://www.esa.int/esaTE/SEMN6HQJNVE_index_0.html, May 2010.

[För06] M. Förtsch, H. Zimmermann, and H. Pless, "220-MHz Monolithically Integrated Optical Sensor With Large-Area Integrated PIN Photodiode", *IEEE Sensors Journal*, vol. 6, no. 2, pp. 385–390, April 2006.

[Gal03] S. Galal and B. Razavi, "10-Gb/s Limiting Amplifier and Laser/Modulator Driver in 0.18-μm CMOS Technology", *IEEE Journal of Solid-State Circuits*, vol. 38, no. 12, pp. 2138–2146, December 2003.

[Gal04] S. Galal and B. Razavi, "40-Gb/s Amplifier and ESD Protection Circuit in 0.18-μm CMOS Technology", *IEEE Journal of Solid-State Circuits*, vol. 39, no. 12, pp. 2389–2396, December 2004.

[Gen01] J. Genoe, D. Coppée, J.H. Stiens, R.A. Vounckx, and M. Kuijk, "Calculation of the Current Response of the Spatially Modulated Light CMOS Detector", *IEEE Transactions on Electron Devices*, vol. 48, no. 9, pp. 1892–1902, September 2001.

[God08] P. Godoy and J.L. Dawson, "Chopper Stabilization of Analog Multipliers, Variable Gain Amplifiers, and Mixers", *IEEE Journal of Solid-State Circuits*, vol. 43, no. 10, pp. 2311–2321, October 2008.

[Gra84] P.R. Gray and R.G. Meyer, *Analysis and Design of Analog Integrated Circuits*, John Wiley & Sons, Inc., 605 Third Avenue, New York, N.Y. 10158, second edition, 1984.

[Hag91] K. Hagimoto, Y. Miyagawa, Y. Miyamoto, M. Ohhata, T. Suzuki, and H. Kikuchi, "Over 10 Gb/s Regenerators Using Monolithic IC's for Lightwave Communication Systems", *IEEE Journal on Selected Areas in Communications*, vol. 9, no. 5, pp. 673–682, June 1991.

[Her05] C. Hermans and M. Steyaert, "A 3.5Gbit/s Post-Amplifier in 0.18μm CMOS", in *Proceedings of the 2005 European Solid-State Circuits Conference*, pp. 431–434, September 2005.

[Her07] C. Hermans and M. Steyaert, *Broadband Opto-Electrical Receivers in Standard CMOS*, Springer, 2007.

[Hua04] L.-R. Huang, C.-M. Tsai, C.-Y. Chien, C.-F. Chang, and D.-U. Lee, "Chip Set Design for 10GB/s Optical Transceiver", in *Proceedings of the 2004 IEEE Asia-Pacific Conference on Circuits and Systems*, volume 1, pp. 113–116, December 2004.

[Hua07] H.-Y. Huang, J.-C. Chien, and L.-H. Lu, "A 10-Gb/s Inductorless CMOS Limiting Amplifier With Third-Order Interleaving Active Feedback", *IEEE Journal of Solid-State Circuits*, vol. 42, no. 5, pp. 1111–1120, May 2007.

[IBM10] IBM, "Silicon Integrated Nanophotonics, development of on-chip optical interconnects for future multi-core processors", http://domino.research.ibm.com/comm/research_projects.nsf/pages/photonics.index.html, May 2010.

[Ina00] U.S. Inan and A.S. Inan, *Electromagnetic waves*, Prentice-Hall, Inc., Upper Saddle River, NJ 07458, 2000.

[Inc10] Virginia Semiconductor Inc., "Optical Properties of Silicon", http://www.virginiasemi.com/pdf/Optical%20Properties%20of%20Silicon71502.pdf, June 2010.

[Ing99] M. Ingels and M.S.J. Steyaert, "A 1-Gb/s, 0.7-μm CMOS Optical Receiver with Full Rail-to-Rail Output Swing", *IEEE Journal of Solid-State Circuits*, vol. 34, no. 7, pp. 971–977, July 1999.

[Ing04] M. Ingels and M. Steyaert, *Integrated CMOS Circuits for Optical Communications*, Springer, 2004.

[Jin87] R.P. Jindal, "Gigahertz-Band High-Gain Low-Noise AGC Amplifiers in Fine-Line NMOS", *IEEE Journal of Solid-State Circuits*, vol. 22, no. 4, pp. 512–521, August 1987.

[Jut05] M. Jutzi, M. Grozing, E. Gaugler, W. Mazioschek, and M. Berroth, "2-Gb/s CMOS Optical Integrated Receiver With a Spatially Modulated Photodetector", *IEEE Photonics Technology Letters*, vol. 17, no. 6, pp. 1268–1270, June 2005.

[Kao09] T.S.-C. Kao and A.C. Carusone, "A 5-Gbps Optical Receiver with Monolithically Integrated Photodetector in 0.18-µm CMOS", in *Proceedings of the 2009 IEEE Custom Integrated Circuits Conference*, pp. 451–454, June 2009.

[Kno03] J. Knorr and H. Zimmermann, "A Transmission-Line Approach for Modeling Feedback Resistors in Integrated Transimpedance Amplifiers", *IEEE Transactions on Circuits and Systems I: Fundamental Theory and Applications*, vol. 50, no. 9, pp. 1192–1195, September 2003.

[Kro05] C. Kromer, G. Sialm, C. Berger, T. Morf, M.L. Schmatz, F. Ellinger, D. Erni, G.-L. Bona, and H. Jackel, "A 100-mW 4x10 Gb/s Transceiver in 80-nm CMOS for High-Density Optical Interconnects", *IEEE Journal of Solid-State Circuits*, vol. 40, no. 12, pp. 2667–2679, December 2005.

[Lee10] D. Lee, J. Han, E. Chang, G. Han, and S.M. Park, "An 8.5Gb/s CMOS OEIC with On-Chip Photodiode for Short-Distance Optical Communications", in *2010 IEEE International Solid-State Circuits Conference, Digest of Technical Papers*, pp. 362–363, February 2010.

[Ler05] P. Leroux and M. Steyaert, *LNA-ESD Co-Design for Fully Integrated CMOS Wireless Receivers*, Springer, 2005.

[Lia06] C.-F. Liao and S.-I. Liu, "A 10Gb/s CMOS AGC Amplifier with 35dB Dynamic Range for 10Gb Ethernet", in *2006 IEEE International Solid-State Circuits Conference, Digest of Technical Papers*, pp. 2092–2101, February 2006.

[Mac10] B. Machiels, P. Reynaert, and M. Steyaert, "Power Efficient Distributed Low-Noise Amplifier in 90 nm CMOS", in *2010 IEEE Radio Frequency Integrated Circuits Symposium, Digest of Papers*, pp. 131–134, May 2010.

[Max10] Maxim, "Statistical Confidence Levels for Estimating Error Probability", http://notes-application.abcelectronique.com /003/3-5321.pdf, October 2010.

[Moo65] G.E. Moore, "Cramming more components onto integrated circuits", *Electronics*, vol. 38, no. 8, pp. 114–117, April 1965.

[Nar07] A. Narasimha, B. Analui, Y. Liang, T.J. Sleboda, S. Abdalla, E. Balmater, S. Gloeckner, D. Guckenberger, M. Harrison, R.G.M.P. Koumans, D. Kucharski, A. Mekis, S. Mirsaidi, Dan Song, and T. Pinguet, "A Fully Integrated 4x10-Gb/s DWDM Optoelectronic Transceiver Implemented in a Standard 0.13 µm CMOS SOI Technology", *IEEE Journal of Solid-State Circuits*, vol. 42, no. 12, pp. 2736–2744, December 2007.

[Nog06] M. Nogawa, Y. Ohtomo, S. Kimura, K. Nishimura, T. Kawamura, and M. Togashi, "A 10Gb/s Burst-Mode Adaptive Gain Select Limiting Amplifier in 0.13µm CMOS", in *2006 IEEE International Solid-State Circuits Conference, Digest of Technical Papers*, pp. 940–949, February 2006.

[Olv92] A.D. Olver, *Microwave and Optical Transmission*, John Wiley & Sons Ltd, Baffins Lane, Chichester, West Sussex PO191UD, England, 1992.

[Pal08] S. Palermo, A. Emami-Neyestanak, and M. Horowitz, "A 90 nm CMOS 16 Gb/s Transceiver for Optical Interconnects", *IEEE Journal of Solid-State Circuits*, vol. 43, no. 5, pp. 1235–1246, May 2008.

[Pel89] M.J.M. Pelgrom, A.C.J. Duinmaijer, and A.P.G. Welbers, "Matching Properties of MOS Transistors", *IEEE Journal of Solid-State Circuits*, vol. 24, no. 5, pp. 1433–1439, October 1989.

[Pis05] M. Pisati, C. Bazzani, F. Gatta, F. Svelto, and R Castello, "A 0.18 µm CMOS TIA Plus Limiting Amplifier with a 5.3 GHz Overall Bandwidth for Fiber Optic Communications", *Analog Integrated Circuits and Signal Processing*, vol. 45, no. 2, pp. 115–122, November 2005.

[POF10] POF+, "Plastic Optical Fibre for Pervasive Low-cost Ultra-high capacity Systems", http://www.ict-pof-plus.eu, February 2010.

[Rab03] J.M. Rabaey, A. Chandrakasan, and B. Nikolić, *Digital Integrated Circuits*, Pearson Education International, Upper Saddle River, NJ 07458, second edition, 2003.

[Rad03] S. Radovanović, A.-J. Annema, and B. Nauta, "Physical and electrical bandwidths of integrated photodiodes in standard CMOS technology", in *Proceedings of the 2003 Conference on Electron Devices and Solid-State Circuits*, pp. 95–98, December 2003.

[Rad05] S. Radovanović, A.-J. Annema, and B. Nauta, "A 3-Gb/s Optical detector in Standard CMOS for 850-nm Optical Communication", *IEEE Journal of Solid-State Circuits*, vol. 40, no. 8, pp. 1706–1717, August 2005.

[Rad06] S. Radovanović, A.-J. Annema, and B. Nauta, *High-Speed Photodiodes in Standard CMOS Technology*, Springer, PO Box 17, 3300 AA Dordrecht, The Netherlands, 2006.

[Raz01] B. Razavi, *Design of Analog CMOS Integrated Circuits*, McGraw-Hill, 1221 Avenue of the Americas, New York, NY 10020, 2001.

[Raz03] B. Razavi, *Design of Integrated Circuits for Optical Communications*, McGraw-Hill, 1221 Avenue of the Americas, New York, NY 10020, 2003.

[Roo00] C. Rooman, D. Coppee, and M. Kuijk, "Asynchronous 250-Mb/s Optical Receivers with Integrated Detector in Standard CMOS Technology for Optocoupler Applications", *IEEE Journal of Solid-State Circuits*, vol. 35, no. 7, pp. 953–958, July 2000.

[Säc00] E. Säckinger and W.C. Fischer, "A 3-GHz 32-dB CMOS Limiting Amplifier for SONET OC-48 Receivers", *IEEE Journal of Solid-State Circuits*, vol. 35, no. 12, pp. 1884–1888, December 2000.

[Säc05] E. Säckinger, *Broadband Circuits for Optical Fiber Communication*, John Wiley & Sons, Inc., Hoboken, New Jersey, 2005.

[Saf08] N. Safavian, K.S. Karim, A. Nathan, and J.A. Rowlands, "A 3-TFT Hybrid Active-Passive Pixel with Correlated Double Sampling CMOS Readout Circuit for Real-Time Medical X-ray Imaging", in *Proceedings of the 2008 European Solid-State Circuits Conference*, pp. 122–125, September 2008.

[San03] M.A.T. Sanduleanu and E. Stikvoort, "Inductor-less, 10Gb/s Limiter with 10mV Sensitivity and Offset/Temperature Compensation in Baseline CMOS18", in *Proceedings of the 2003 European Solid-State Circuits Conference*, pp. 153–156, September 2003.

[San06] W.M.C. Sansen, *Analog Design Essentials*, Springer, PO Box 17, 3300 AA Dordrecht, The Netherlands, 2006.

[Sch02] K. Schrodinger, J. Stimma, and M. Mauthe, "A Fully Integrated CMOS Receiver Front-End for Optic Gigabit Ethernet", *IEEE Journal of Solid-State Circuits*, vol. 37, no. 7, pp. 874–880, July 2002.

[Sch09] C.L. Schow, F.E. Doany, C. Chen, A.V. Rylyakov, C.W. Baks, D.M. Kuchta, R.A. John, and J.A. Kash, "Low-Power 16 x 10 Gb/s Bi-Directional Single Chip CMOS Optical Transceivers Operating at < 5 mW/Gb/s/link", *IEEE Journal of Solid-State Circuits*, vol. 44, no. 1, pp. 301–313, January 2009.

[Sei03] C. Seidl, J. Knorr, and H. Zimmermann, "Simple feedback network for bandwidth enhancement of transimpedance amplifiers", *Electronics Letters*, vol. 39, no. 25, pp. 1849–1851, November 2003.

[Sei04] C. Seidl, J. Knorr, and H. Zimmermann, "Single-Stage 378MHz 178kΩ Transimpedance Amplifier with Capacitive-Coupled Voltage Dividers", in *2004 IEEE International Solid-State Circuits Conference, Digest of Technical Papers*, pp. 470–540, February 2004.

[Stu05] J. Sturm, M. Leifhelm, H. Schatzmayr, S. Groiss, and H. Zimmermann, "Optical Receiver IC for CD/DVD/Blue-Laser Application", *IEEE Journal of Solid-State Circuits*, vol. 40, no. 7, pp. 1406–1413, July 2005.

[Swo06] R. Swoboda and H. Zimmermann, "11Gb/s Monolithically Integrated Silicon Optical Receiver for 850nm Wavelength", in *2006 IEEE International Solid-State Circuits Conference, Digest of Technical Papers*, pp. 904–911, February 2006.

[Tad08] N. Tadić and H. Zimmermann, "Optical Receiver With Widely Tunable Sensitivity in BiCMOS Technology", *IEEE Transactions on Circuits and Systems I: Regular Papers*, vol. 55, no. 5, pp. 1223–1236, June 2008.

[Tak98] T. Takimoto, N. Fukunaga, M. Kubo, and N. Okabayashi, "High Speed Si-OEIC (OPIC) for Optical Pickup", *IEEE Transactions on Consumer Electronics*, vol. 44, no. 1, pp. 137–142, February 1998.

[Tan98] A. Tanabe, M. Soda, Y. Nakahara, T. Tamura, K. Yoshida, and A. Furukawa, "A Single-Chip 2.4-Gb/s CMOS Optical Receiver IC with Low Substrate Cross-Talk Preamplifier", *IEEE Journal of Solid-State Circuits*, vol. 33, no. 12, pp. 2148–2153, December 1998.

[Tsa04] C.-M. Tsai, "A 20mW 85dBΩ 1.25Gb/s CMOS Transimpedance Amplifier with Photodiode Capacitance Cancellation", in *2004 Symposium on VLSI Circuits, Digest of Technical Papers*, pp. 408–409, June 2004.

[Tsa05] C.-M. Tsai and L.-R. Huang, "A 21 mW 2.5Gb/s 15kΩ Self-Compensated Differential Transimpedance Amplifier", in *2005 IEEE International Solid-State Circuits Conference, Digest of Technical Papers*, pp. 234–595, February 2005.

[Tsa09] C.-M. Tsai, "A 40mW 3Gb/s Self-Compensated Differential Transimpedance Amplifier With Enlarged Input Capacitance Tolerance in 0.18 μm CMOS Technology", *IEEE Journal of Solid-State Circuits*, vol. 44, no. 10, pp. 2671–2677, October 2009.

[VO00] R. Van Overstraeten and P. Heremans, *Semiconductor Devices*, Acco, Leuven, Belgium, 2000.

[Wei06] J.R.M. Weiss, M.A. Kossel, C. Menolfi, T. Morf, M.L. Schmatz, T. Toifl, and H. Jaeckel, "A DC-to-44-GHz 19dB Gain Amplifier in 90nm CMOS Using Capacitive Bandwidth Enhancement", in *2006 IEEE International Solid-State Circuits Conference, Digest of Technical Papers*, pp. 2082–2091, February 2006.

[Wit07] J.F. Witte, K.A.A. Makinwa, and J.H. Huijsing, "A CMOS Chopper Offset-Stabilized Opamp", *IEEE Journal of Solid-State Circuits*, vol. 42, no. 7, pp. 1529–1535, July 2007.

[Wit09] J.F. Witte, J.H. Huijsing, and K.A.A. Makinwa, "A Chopper and Auto-Zero Offset-Stabilized CMOS Instrumentation Amplifier", in *2009 Symposium on VLSI Circuits, Digest of Technical Papers*, pp. 210–211, June 2009.

[Woo98] T.K. Woodward and A.V. Krishnamoorthy, "1Gbit/s CMOS photoreceiver with integrated detector operating at 850nm", *Electronics Letters*, vol. 34, no. 12, pp. 1252–1253, June 1998.

[Yoo07] K. Yoo, D. Lee, G. Han, S.M. Park, and W.S. Oh, "A 1.2V 5.2mW 40dB 2.5Gb/s Limiting Amplifier in 0.18μm CMOS Using Negative-Impedance Compensation", in *2007 IEEE International Solid-State Circuits Conference, Digest of Technical Papers*, pp. 56–57, February 2007.

[You10] I.A. Young, E. Mohammed, J.T.S. Liao, A.M. Kern, S. Palermo, B.A. Block, M.R. Reshotko, and P.L.D. Chang, "Optical I/O Technology for Tera-Scale Computing", *IEEE Journal of Solid-State Circuits*, vol. 45, no. 1, pp. 235–248, January 2010.

[Yun09] J.S. Yun, M. Seo, B. Choi, J. Han, Y. Eo, and S.M. Park, "A 4Gb/s Current-Mode Optical Transceiver in 0.18μm CMOS", in *2009 IEEE International Solid-State Circuits Conference, Digest of Technical Papers*, pp. 102–103, February 2009.

[Zim04] H. Zimmermann, *Silicon Optoelectronic Integrated Circuits*, Springer, 2004.

Index

F. Tavernier, M. Steyaert, *High-Speed Optical Receivers with Integrated Photodiode in Nanoscale CMOS,* Analog Circuits and Signal Processing 5,
DOI 10.1007/978-1-4419-9925-2, © Springer Science+Business Media, LLC 2011